Roger Lee (Ed.)

Software Engineering Research, Management and Applications 2011

Studies in Computational Intelligence, Volume 377

Editor-in-Chief
Prof. Janusz Kacprzyk
Systems Research Institute
Polish Academy of Sciences
ul. Newelska 6
01-447 Warsaw
Poland
E-mail: kacprzyk@ibspan.waw.pl

Further volumes of this series can be found on our homepage: springer.com

Vol. 354. Ljupco Kocarev and Shiguo Lian (Eds.)
Chaos-Based Cryptography, 2011
ISBN 978-3-642-20541-5

Vol. 355. Yan Meng and Yaochu Jin (Eds.)
Bio-Inspired Self-Organizing Robotic Systems, 2011
ISBN 978-3-642-20759-4

Vol. 356. Slawomir Koziel and Xin-She Yang (Eds.)
Computational Optimization, Methods and Algorithms, 2011
ISBN 978-3-642-20858-4

Vol. 357. Nadia Nedjah, Leandro Santos Coelho, Viviana Cocco Mariani, and Luiza de Macedo Mourelle (Eds.)
Innovative Computing Methods and their Applications to Engineering Problems, 2011
ISBN 978-3-642-20957-4

Vol. 358. Norbert Jankowski, Włodzisław Duch, and Krzysztof Grąbczewski (Eds.)
Meta-Learning in Computational Intelligence, 2011
ISBN 978-3-642-20979-6

Vol. 359. Xin-She Yang, and Slawomir Koziel (Eds.)
Computational Optimization and Applications in Engineering and Industry, 2011
ISBN 978-3-642-20985-7

Vol. 360. Mikhail Moshkov and Beata Zielosko
Combinatorial Machine Learning, 2011
ISBN 978-3-642-20994-9

Vol. 361. Vincenzo Pallotta, Alessandro Soro, and Eloisa Vargiu (Eds.)
Advances in Distributed Agent-Based Retrieval Tools, 2011
ISBN 978-3-642-21383-0

Vol. 362. Pascal Bouvry, Horacio González-Vélez, and Joanna Kolodziej (Eds.)
Intelligent Decision Systems in Large-Scale Distributed Environments, 2011
ISBN 978-3-642-21270-3

Vol. 363. Kishan G. Mehrotra, Chilukuri Mohan, Jae C. Oh, Pramod K. Varshney, and Moonis Ali (Eds.)
Developing Concepts in Applied Intelligence, 2011
ISBN 978-3-642-21331-1

Vol. 364. Roger Lee (Ed.)
Computer and Information Science, 2011
ISBN 978-3-642-21377-9

Vol. 365. Roger Lee (Ed.)
Computers, Networks, Systems, and Industrial Engineering 2011, 2011
ISBN 978-3-642-21374-8

Vol. 366. Mario Köppen, Gerald Schaefer, and Ajith Abraham (Eds.)
Intelligent Computational Optimization in Engineering, 2011
ISBN 978-3-642-21704-3

Vol. 367. Gabriel Luque and Enrique Alba
Parallel Genetic Algorithms, 2011
ISBN 978-3-642-22083-8

Vol. 368. Roger Lee (Ed.)
Software Engineering, Artificial Intelligence, Networking and Parallel/Distributed Computing 2011, 2011
ISBN 978-3-642-22287-0

Vol. 369. Dominik Ryżko, Piotr Gawrysiak, Henryk Rybinski, and Marzena Kryszkiewicz (Eds.)
Emerging Intelligent Technologies in Industry, 2011
ISBN 978-3-642-22731-8

Vol. 370. Alexander Mehler, Kai-Uwe Kühnberger, Henning Lobin, Harald Lüngen, Angelika Storrer, and Andreas Witt (Eds.)
Modeling, Learning, and Processing of Text Technological Data Structures, 2011
ISBN 978-3-642-22612-0

Vol. 371. Leonid Perlovsky, Ross Deming, and Roman Ilin (Eds.)
Emotional Cognitive Neural Algorithms with Engineering Applications, 2011
ISBN 978-3-642-22829-2

Vol. 372. António E. Ruano and Annamária R. Várkonyi-Kóczy (Eds.)
New Advances in Intelligent Signal Processing, 2011
ISBN 978-3-642-11738-1

Vol. 373. Oleg Okun, Giorgio Valentini, and Matteo Re (Eds.)
Ensembles in Machine Learning Applications, 2011
ISBN 978-3-642-22909-1

Vol. 374. Dimitri Plemenos and Georgios Miaoulis (Eds.)
Intelligent Computer Graphics 2011, 2011
ISBN 978-3-642-22906-0

Vol. 375. Marenglen Biba and Fatos Xhafa (Eds.)
Learning Structure and Schemas from Documents, 2011
ISBN 978-3-642-22912-1

Vol. 376. Toyohide Watanabe and Lakhmi C. Jain (Eds.)
Innovations in Intelligent Machines – 2, 2012
ISBN 978-3-642-23189-6

Vol. 377. Roger Lee (Ed.)
Software Engineering Research, Management and Applications 2011, 2012
ISBN 978-3-642-23201-5

Roger Lee (Ed.)

Software Engineering Research, Management and Applications 2011

Springer

Editor

Prof. Roger Lee
Central Michigan University
Computer Science Department
Software Engineering & Information
Technology Institute
Mt. Pleasant, MI 48859
USA
E-mail: lee1ry@cmich.edu

ISBN 978-3-642-27091-8 ISBN 978-3-642-23202-2 (eBook)

DOI 10.1007/978-3-642-23202-2

Studies in Computational Intelligence ISSN 1860-949X

Typeset & Cover Design: Scientific Publishing Services Pvt. Ltd., Chennai, India.

Printed on acid-free paper

9 8 7 6 5 4 3 2 1

springer.com

Preface

The purpose of the 9th International Conference on Software Engineering Research, Management and Applications(SERA 2011) held on August 10–12, 2011 Baltimore, Maryland was to bring together researchers and scientists, businessmen and entrepreneurs, teachers and students to discuss the numerous fields of computer science, and to share ideas and information in a meaningful way. Our conference officers selected the best 13 papers from those papers accepted for presentation at the conference in order to publish them in this volume. The papers were chosen based on review scores submitted by members of the program committee, and underwent further rounds of rigorous review.

In Chapter 1, Garcia Ivan et al. The main achievement of this paper, lies in the development of an integrated mechanism for assessing software processes, using a hybrid mechanism that incorporates modeling-based assessment. This mechanism was evaluated using the EvalProSoft framework and descriptive concepts, to facilitate establishing SPI initiatives in a small Mexican software company.

In Chapter 2, Dan Cramer et al. In this paper, we develop a model that will help policy makers anticipate the occurrences of emergencies. Spatial analysis methods such as hotspot analysis are used that can help policy makers distribute resources fairly by needs.

In Chapter 3, Kwangchun Lee et al. In this paper, we propose a market-driven quantitative scoping method. This method incorporates customers needs, product family structure and market strategies into scoping such that this ensures that SPL derivatives penetrate market grids.

In Chapter 4, Kazunori Iwata et al. In this paper, we cluster and analyze data from the past embedded software development projects using self-organizing maps (SOMs)[9] that are a type of artificial neural networks that rely on unsupervised learning. The purpose of the clustering and analysis is to improve the accuracy of predicting the number of errors. A SOM produces a low-dimensional, discretized representation of the input space of training samples; these representations are called maps. SOMs are useful for visualizing low-dimensional views of high dimensional data, a multidimensional scaling technique. The advantages of SOMs for statistical applications are as follows: (1) data visualization, (2) information processing on association and recollection, (3) summarizing large-scale data, and (4) creating nonlinear models. To verify our approach, we perform an evaluation experiment that compares SOM classification to product type classification using Welch's t-test for Akaike's Information Criterion (AIC). The results indicate that the SOM classification method is more contributive than product type classification in creating estimation models, because the mean AIC of SOM classification is statistically significantly lower.

In Chapter 5, Seonah Lee et al. While performing an evolution task, programmers spend significant time trying to understand a code base. To facilitate programmers' comprehension of code, researchers have developed software visualization tools. However, those tools have not predicted the information that programmers seek during their program comprehension activities. To responsively provide informative diagrams in a timely manner, we suggest a graphical code recommender and conduct an iterative Wizard of Oz study in order to examine when and what diagrammatic contents should appear in a graphical view to guide a programmer in exploring source code. We found that programmers positively evaluate a graphical code recommender that changes in response to their code navigation. They favored a graphical view that displays the source locations frequently visited by other programmers during the same task. They commented that the graphical code recommender helped in particular when they were uncertain about where to look while exploring the code base.

In Chapter 6, Haeng-Kon Kim and Roger Y. Lee. This survey report is the analysis of the model based regression testing techniques according to the parameter identified during this study. The summary as well as the analysis of the approaches is discussed in this survey report. In the end we concluded the survey by identifying the areas of further research in the field of model based regression testing.

In Chapter 7, Frederik Schmidt et al. This paper describes the initial development of a framework for automatic software architecture reconstruction and source code migration. This frame-work offers the potential to reconstruct the conceptual architecture of software systems and to automatically migrate the physical architecture of a software system toward a conceptual architecture model. The approach is implemented within a proof of concept prototype which is able to analyze java system and reconstruct a conceptual architecture for these systems as well as to re-factor the system towards a conceptual architecture.

In Chapter 8, Haeng-Kon Kim. In this paper, we discuss the creation of such a model and its relevance for technical design of a smart agent for u-learning mobile software system. Conventional approaches to modeling of context focus either on the application domain or the problem domain. These approaches are presented and their relevance for technical design and modeling of software for agent mobile systems is discussed. The paper also reports from an empirical study where a methodology that combines both of these approaches was introduced and employed for modeling of the domain-dependent aspects that were relevant for the design of a software component for mobile agents. We also discuss some pertinent issues concerning the deployment of intelligent agents on mobile devices for certain interaction paradigms are discussed and illustrated in the context of a u-learning applications.

In Chapter 9, Raza Hasan et al. This paper investigates software maintenance practices in a small information systems organization to come up with the nature and categories of heuristics used that successfully guided the software maintenance process. Specifically, this paper documents a set of best practices that small organizations can adopt to facilitate their software maintenance processes in the absence of maintenance-specific guidelines based on preliminary empirical investigation.

In Chapter 10, Hui Liu et al. Cell phones are among the most common types of technologies present today and have become an integral part of our daily activities. The latest statistics indicate that currently there are over five billion mobile subscribers are in the world and increasingly cell phones are used in criminal activities and confiscated at the crime scenes. Data extracted from these phones are presented as evidence in the court, which has made digital forensics a critical part of law enforcement and legal systems in the world. A number of forensics tools have been developed aiming at extracting and acquiring the ever-increasing amount of data stored in the cell phones; however, one of the main challenges facing the forensics community is to determine the validity, reliability and effectiveness of these tools. To address this issue, we present the performance evaluation of several market- leading forensics tools in the following two ways: the first approach is based on a set of evaluation standards provided by National Institute of Standards and Technology (NIST), and the second approach is a simple and effective anti-forensics technique to measure the resilience of the tools.

In Chapter 11, Je-Kuk Yun et al. In this paper, we present a simple but feasible scenario that helps a package receiver simply track the current location of its delivery truck for the package.

In Chapter 12, Mohamad Kassab et al. In this paper, we propose an approach for a quantitative evaluation of the support provided by a pattern for a given targeted set of quality attributes.

In Chapter 13, Purushothaman Surendran et al. The objective of this paper is to analyze the detection performance of non-coherent detectors. The non-coheron detectors have almost similar detection probability in a background of white Gaussian noise. The performance of the detector is analyzed and simulation has been done in order to verify.

It is our sincere hope that this volume provides stimulation and inspiration, and that it will be used as a foundation for works yet to come.

August 2011 Guest Editors

Yeong-Tae Song
Chao Lu

Contents

List of Contributors

Subrata Acharya
Towson University, USA
sacharya@towson.edu

Yoshiyuki Anan
Base Division, Omron Software Co., Ltd., Japan
y-anan@mx.omronsoft.co.jp

Shiva Azadegan
Towson University, USA
azadegan@towson.edu

Albert Arthur Brown
Central Michigan University, USA

Changhyun Byun
Towson University, USA
cbyun1@towson.edu

J.A. Calvo-Manzano
Technical University of Madrid
jacalvo@fi.upm.es

Suranjan Chakraborty
Towson University, USA
schakraborty@towson.edu

Juno Chang
Sangmyung University, Korea
jchang@smu.ac.kr

Andrew M. Connor
Auckland University, New Zealand
Andrew.connor@aut.ac.nz

Dan Cramer
Central Michigan University, USA

D. Cruz
Cañada University, Canada
dago@naxoloxa.unca.edu.mx

Josh Dehlinger
Towson University, USA
jdehlinger@towson.edu

Ghizlane El-Boussaidi
École de technologie supérieure, Québec
ghizlane.elboussaidi@etsmtl.ca

I. Garcia
Technological University of the Mixtec Region
ivan@mixteco.utm.mx

Raza Hasan
Towson University, USA
rhasan@towson.edu

Gongzhu Hu
Central Michigan University, USA
hu1g@cmich.edu

Naohiro Ishii
Aichi Institute of Technology, Japan
ishii@aitech.ac.jp

Kazunori Iwata
Aichi University, Japan
kazunori@vega.aichi-u.ac.jp

Sungwon Kang
Department of Computer Science, KAIST
sungwon.kang@kaist.ac.kr

Mohamad Kassab
Concordia University, Quebec.
moh_kass@cs.concordia.ca

Haeng-Kon, Kim
University of Daegu, Korea
hangkon@cu.ac.kr

Yanggon Kim
Towson University, USA
ykim@towson.edu

Seok Jun Ko
Jeju National University, Jeju, Korea
sjko@jejunu.ac.kr

Dan Hyung Lee
Korea Advanced Institute of Science and Technology, Korea
danlee@kaist.ac.kr

Jong-Hun Lee
DGIST, Daegu, Korea

Kwangchun Lee
Korea Advanced Institute of Science and Technology, Korea
statkclee@kaist.ac.kr

Roger Y. Lee
Central Michigan University, USA
lee1ry@cmich.edu

Seonah Lee
Department of Computer Science, KAIST
saleese@kaist.ac.kr

Hui Liu
Towson University, USA
janetliuhui@gmail.com

Stephen G. MacDonell
Auckland University of Technology, New Zealand
stephen.macdonell@aut.ac.nz

Hafedh Mili
Université du Québec à Montréal, Canada
hafedh.mili@uqam.ca

Toyoshiro Nakashima
Sugiyama Jogakuen University, Japan
nakasima@sugiyama-u.ac.jp

C. Pacheco
Technological University of the Mixtec Region
leninca@mixteco.utm.mx

Kyungeun Park
Towson University, USA
kpark3@towson.edu

Frederik Schmidt
Auckland University of Technology, New Zealand
fschmidt@aut.ac.nz

Ali Sistani
Towson University, USA
masistani@yahoo.com

Purushothaman Surendran
Jeju National University, Jeju, Korea

Wei Yu
Towson University, USA
wyu@towson.edu

Je-Kuk Yun
Towson University, USA
jyun4@towson.edu

Implementing the Modeling-Based Approach for Supporting the Software Process Assessment in SPI Initiatives Inside a Small Software Company

I. Garcia, C. Pacheco, D. Cruz, and J.A. Calvo-Manzano

Abstract. Software Process Improvement (SPI) has become more and more important during the past ten years, since competition is increasingly determined by the proportion of software products and services. Over the years, many different SPI approaches have been developed; most of them are either best-practice-oriented approaches or continuous improvement approaches, which require an accurate assessment process as a basis from which to begin the improvement. Usually, almost all research is focused on a questionnaire-based approach for process' assessment. However, without the organizational commitment it is too difficult obtain realistic and accurate results. The main achievement of this paper, lies in the development of an integrated mechanism for assessing software processes, using a hybrid mechanism that incorporates modeling-based assessment. This mechanism was evaluated using the EvalProSoft framework and descriptive concepts, to facilitate establishing SPI initiatives in a small Mexican software company.

1 Introduction

Software has increasingly penetrated society during the last decades. Software is not only used as an independent product by itself, but also as part of everyday systems such as kitchen appliances, washing machines, or mobile phones. However, the increasing significance of software in our daily lives leads to an increasing

I. Garcia · C. Pacheco
Technological University of the Mixtec Region
e-mail: {ivan,leninca}@mixteco.utm.mx

D. Cruz
Cañada University
e-mail: dago@naxoloxa.unca.edu.mx

J.A. Calvo-Manzano
Technical University of Madrid
e-mail: jacalvo@fi.upm.es

R. Lee (Ed.): Software Eng. Research, Management & Appl. 2011, SCI 377, pp. 1–13.
springerlink.com

potential for risks, because the effects of software failures are continuously increasing as well [22]. To respond to this, more and more attention must also be paid to systematic software development and proper quality assurance strategies. Nevertheless, the topic of quality assurance in software development has been neglected for quite a long time. The reason for this is that software is mostly an individual product rather than a mass product. This means that in contrast to traditional production, the focus of quality assurance has to be set during the development process and not during the production process. In this context, with the importance of software and software quality, the community of software processes has grown as well. Since the first International Software Process Workshop (ISPW) held in 1984, when Watts Humphrey stated that *"the quality of the software development process is in close relationship with the quality of the resulting software"* [11]; several process assessment methods have been developed (i.e., CBA-IPI [6], SCAMPI [23], ISO/IEC 15504:2004 [14]), and a large number of SPI approaches, including technologies, (i.e., SPICE [13], ISO/IEC 15504:2004 [14], CMMI-DEV v1.2 [5], ISO/IEC 12207:2008 [12]) exist and are used in industry today. The Humphrey' book "Managing the Software Process" started with the following two statements: an ancient Chinese proverb *"if you don't know where you're going any road will do"* and the Humphrey' version of this: *"if you don't know where you are a map won't help"*. They both seem to be very trivial and easy to understand, but transferring these statements to the software engineering world presents the following three questions, which should be addressed by any process improvement activity:

1. How good is my current software process?.
2. What must I do to improve it?.
3. Where do I start?.

What does the first question mean? It means obtaining a baseline of the current development activities. This idea was supported by the original framework used by SEI's Software Engineering Process Management (SEPM) Program, to characterize its SPI mission, which can be explained in terms of what, how, and why perspectives. A *what* question is related to software process assessments and might then be the following: Do you establish a controlled project plan for software development as a whole? The answer to this question is that it is used as a process appraisal mechanism to establish how an organization should plan the development work. Thus, software process assessments are typically the first step to commencing the SPI.

In order to ensure and facilitate the implementation of assessment findings, an accepted process description reflecting the current status is needed. This means that an integration of process assessment and process modeling, is vital to ensure successful and sustainable software process improvement activities. However, modeling is used only after the assessment process has finished for establishing or depicting existing and renewed software processes. We think that this idea can be applied from when the assessment begins. This research presents a lightweight mechanism which integrates modeling-based and questionnaire-based assessments. This mechanism unifies an existing questionnaire-based assessment method, and an adapted, modeling-based method. Our assessment mechanism has a low resource overhead and does not dictate either a plan-driven or agile process improvement model, making it an attractive assessment method for small software companies. The rest of the paper is structured as follows: Section 2 discusses related work; Section 3 describes the proposed hybrid-mechanism and development of the modeling component; and Section 4 describes the implementation of the modeling-based assessment; with an existing two-phase questionnaire and their application with the Software Industry Process Model (MoProSoft), and some discussion related to the obtained results. Finally, conclusions are shown in Section 5.

2 Related Work

Small software companies find that many assessment methods are linked to plan-driven improvement models and modeling tools that can be expensive in terms of the resources required [19]. We now see cases where high-maturity ratings do not always result in the related processes being used in the subsequent projects. It is because, sometimes, the appraisal process is faulty or organizations are dishonest, and a maturity framework does not focus on how the work is actually done; it only addresses what is done. To avoid these problems, some research has been done.

Research by [27], for example, presents the MR-MPS Process Reference Model and the MA-MPS Process Assessment Method. They were created according to the Brazilian reality in the MPS.BR Project [26], aiming at improving software process mainly in small to medium-size enterprises. They are compatible with CMMI [4] and conformant with ISO/IEC 15504 [14] and ISO/IEC 12207 [12]. McCaffery et al. [19] support the necessity of an alternative assessment method with statistics from the Irish software industry and present a lightweight assessment method called Adept. Adept unifies an existing plan-driven assessment method and a riskbased Agility/Discipline Assessment method. In [25], the authors describe the Improvement Ability (ImprovAbility) Model to measure an organization's or project's ability to succeed with improvement. The method for gathering information during an assessment is inspired primarily by the Bootstrap method [16]. After a group interview, the assessors answer a questionnaire in a spreadsheet form which generates a picture of the strong and weak parameters on a scale from zero to 100. Research on questionnaire-based assessment has dominated the efforts to improve the way to obtain a snapshot of the organization's software practices; for example, the SEI' Maturity Questionnaire [28], which was developed for the SW-CMM model. This maturity questionnaire focused on the maturity of the process without paying attention to finding a weakness in the practices. Using the maturity questionnaire limits the information to two extreme ends: *Yes*, if the practice is performed and *No* if the practice is not performed. Therefore, it does not leave room for intermediate points. Questionnaires with limited answer options may provide limited or misleading information. For example, a project sponsored by the SEI "CMMI Interpretive Guidance Project" supports this argument [3]. However, in one question of the same project, the SEI used five possible responses: *Almost always*, *More often than not*, *Sometimes*, *Rarely if ever* and *Don't know*. As a result, a wider distribution of the types of responses was obtained. The report does not explain, however, the reasons why this methodology was not used in the same way for specific and generic practice questions.

The report of the Process Improvement Program for the Northrop Grumman Information Technology Company [18], proposed a Questionnaire-Based Appraisal with seven possible responses: *Does Not Apply, Don't know, No, about 25% of the time, about 50% of the time, about 75% of the time*, and *Yes*. This work proposed more response granularity. Research by [1], proposed for the appraisal stage, a questionnaire structure using five types of responses: *Always* when the practice is documented and performed between 100% - 75% of the time; *More often* when the practice is documented and performed between 75% - 50% of the time; *Sometimes* when the practice is not documented and is performed between 50% - 25% of the time; *Rarely* when the practice could be documented or not and it is performed between 25% - 0% of the time; *Never* when the practice is not performed in the organization. The responses granularity is similar to [18] and provides more information about the current state of the practices. Cuevas, Serrano and Serrano [2] proposed an assessment methodology based on a questionnaire to identify which practices of the requirements management process are performed but not documented, which practices require prioritising, and which are not implemented due to bad management or unawareness. The last study reviewed was our questionnaire proposed in [10], to obtain a baseline snapshot of Project Management processes using a two-phase questionnaire; to identify both performed and non-performed practices. The proposed questionnaire is based on the Level 2 process areas of the Capability Maturity Model Integration for Development v1.2.

As we can see, the majority of work is focused on developing questionnaire-based mechanisms to improve the assessment process within an SPI initiative. In this paper, we propose a different mechanism based on the modeling approach to obtain rapid assessment, and combine it with an existing questionnaire-based approach [10] to improve the results.

3 A Hybrid Mechanism for Software Process Assessment

As we said, small software companies find that many assessment methods are linked to plan-driven improvement models and can be expensive in terms of the resources required. Our mechanism unifies an existing questionnaire-based assessment method and an, adapted, modeling-based method. Research is focused on developing an alternative assessment mechanism that consists of two phases: the first, and already used before, is the questionnaire-based assessment that works like an instrument for obtaining data based on the practices suggested by a reference model; the second phase is based on the current development process of software companies obtained through the modeling process. The phase of process modeling is hitherto unexplored in the tools for supporting the software process assessment. The aim of linking these two types of assessments is to obtain a more accurate result because, having two types of assessment mechanisms with different structures and presentation, but based on the same principles, an automated mechanism can more easily detect inconsistencies related to the users' responses. When assessing the current processes in small companies, if the modeling-based assessment determines that a project manager does not perform a certain activity, and later the questionnaire-based assessment shows that such activity is performed within the company, we can infer that the project manager is responding randomly or lies in their responses.

3.1 The Modeling-Based Assessment Mechanism

The process modeling component focuses on the current company's software development process. This component uses the Process Change Methodology proposed by [8] to describe the current software development process of the company. To accomplish this task, the mechanism introduces the use of a graphical notation to edit the current process of the company. In this sense, the SEI in collaboration with the SEPG (OrgName's Software Engineering Process Groups) proposed a graphical notation to construct diagrams that show the phases of the software development process.

In particular, flow charts are used to illustrate the behavioral perspective lifecycle in the software development process. Each phase of the software development process consists of activities and flow charts that can be used to describe the relationships between these activities (as the sequence, placement and decisions). This kind of assessment is based on the execution of phases from the generic lifecycle model and some critical activities during the development process. The mechanism possesses the characteristic restrictions of any editor of flow diagrams to carry out the assessment in an effective way:

- All diagrams should have a beginning symbol.
- The diagram should conclude with a final component.
- The number of activities should be sequential avoiding the jump of numbers.
- The sequence of activities should be orderly.
- Components or a block of components cannot exist, if they are not connected with the general diagram.

Some of these restrictions are shown when carrying out the diagram and others in the stage of results. The mechanism works in a sequential way, looking for those activities proposed as an "ideal diagram" by the reference model (or process model), besides checking that those activities are carried out in the phases that correspond. Taking the sequence of the diagram of a particular process, our mechanism can verify what activities are outside of sequence and the activities that are carried out in phases different to those that belong. Another point to consider is that the activities have a coherent sequence. Likewise, it is verified that those responsible for certain activities fulfill the role that is specified in the reference model, independently of the name that is given in a company to a certain area of work. Also, the modeling mechanism verifies that those responsible for certain activities, perform the role specified in the reference model, regardless of the name given in a firm for a particular job.

As a result of the modeling-based assessment, a new diagram is obtained through a mapping of the two diagrams current (organization) and ideal (process model). This new diagram is composed of the activities in the organization's process that have fulfilled the criteria of an evaluation of the mechanism, as well as the activities of the ideal process that are not found during the mapping process (see Figure 1).

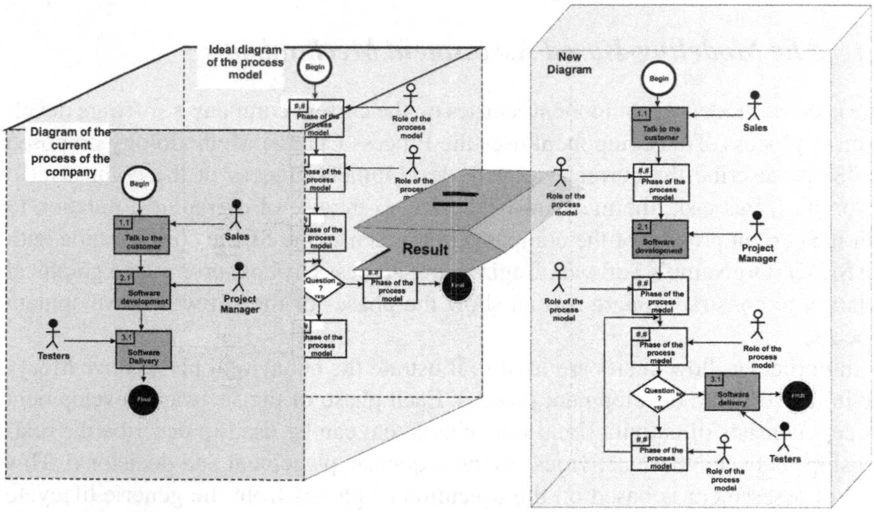

Fig. 1 Mapping the current process and ideal process

The new diagram offers a closer perspective of what should be an ideal process in a company based on the activities that are carried out at that moment. Besides enriching the diagram, the mechanism marks those activities that can be not well located in the organization's process or the roles that could not be responsible for a certain activity. The preconditions to carry out the 'mapping algorithm' execute queries to the database to obtain the elements of the diagram of the current process and the elements of the ideal diagram, both elements will be stored respectively in *currentDiagram* and *idealDiagram*, both tables will be used in the algorithm, besides a new table called *newDiagram*, that will store the elements of the enriched diagram.

3.2 *Combining Questionnaire-Based and Modeling-Based Assessments*

The context of this phase consists of two parts: a mechanism to generate the results of the questionnaire and another to generate the results of the process modeling. The first consists in emptying the database obtained with the answers from the users to the questionnaire-based assessment developed in the assessment phase of an improvement model. Once this data is obtained from the mechanism, based on the qualification guidelines proposed by the assessment model, the results are graphically displayed; the acting level is shown by each assessed process, as well as, the qualifications obtained for each activity.

The second mechanism is responsible for generating the results from the process modeling-based assessment; this mechanism uses the model generated by the user in the assessment phase together with the ideal process previously kept in a database of assets. The results show a new diagram enriched with activities not found in the current process and contrasting with the erroneous activities. In addition, the mistakes and weaknesses, detected in the current organization' process, are provided in the form of a list. With this it provides the user with two forms for analyzing the results of the modeling-based assessment. Figure 2 shows that both mechanisms work together to detect incoherencies in the assessment due to the lack of commitment and veracity of the assessed personnel. Depending on the level of inconsistencies among the two assessments, the qualification of the process or activity will be penalized, or in its defect will be annulled. Finally, this phase provides an estimate of the level of capacity in the processes, according to the limits marked by the reference model.

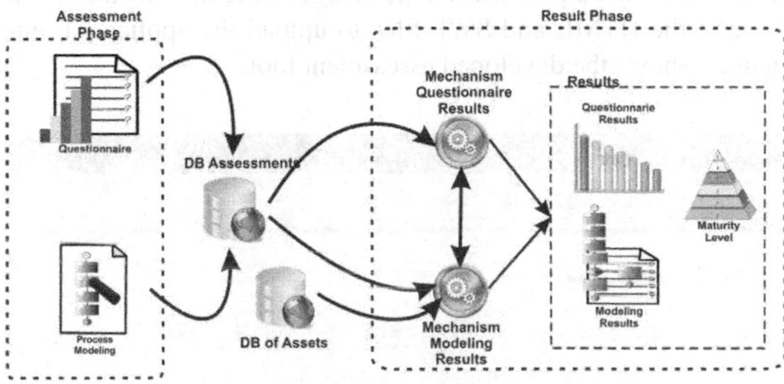

Fig. 2 Context of the work for results phase

4 Implementing the Hybrid Mechanism

To test the hybrid mechanism a computational tool was developed to automate the assessment process, the SelfVation tool. First, we selected a process model focused on small software companies. In Mexico, the MoProSoft model [20] and its process-assessment method (EvalProSoft) were defined [21]. Since the approval of MoProSoft and EvalProSoft as the Mexican standard NMX-I-059-NYCE-2005, the interest of the small companies in acquiring this certification has grown. Besides the certification, as a result of the correlation between the process quality and the obtained product quality, small Mexican companies have gained impetus in improving their software processes, as a strategy to assure the quality of their software products [7]. However, besides the certification desire, the small companies have the problem of adopting quickly and efficiently to the standard NMX-I-059-NYCE-2005 without loss of resources and time to develop new projects. We decided to analyze the

MoProSoft and build an ideal model (diagram) to implement our hybrid mechanism and support SPI initiatives.

Due to the fact that a modeling mechanism requires a lot of dynamic events, this tool was developed, combining the UWE [15] and RUX-Method [17] to develop an RIA (Rich Internet Application) [24]. The UWE provides a specific notation for a graphic representation of Web applications and a method to manage the Web systems development. The RUX method is focused on modeling and managing the RIA. In this research, the UWE is used to specify the content, navigation, and business processes of the Web application; and the RUX method is used to add the typical capabilities of an enriched GUI. The conceptual model includes the objects implied in the interaction between the user and the software tool. The conceptual design develops the model of classes ignoring the navigation paths, presentation and topics related to interaction. The RUX method distinguishes three different levels of interface: abstract interface, concrete interface and final interface. This method uses the Web 1.0 interface previously developed with UWE. Once the design stage was covered, the tool implementation was performed using the ActionScript code and the MXML (Multimedia eXtensible Markup Language) from Adobe FLEX language; PHP files were used to construct the bridge between SelfVation and queries to databases, and the HTML and SWF files to upload the application into a Web browser. Figure 3 shows the developed assessment tool.

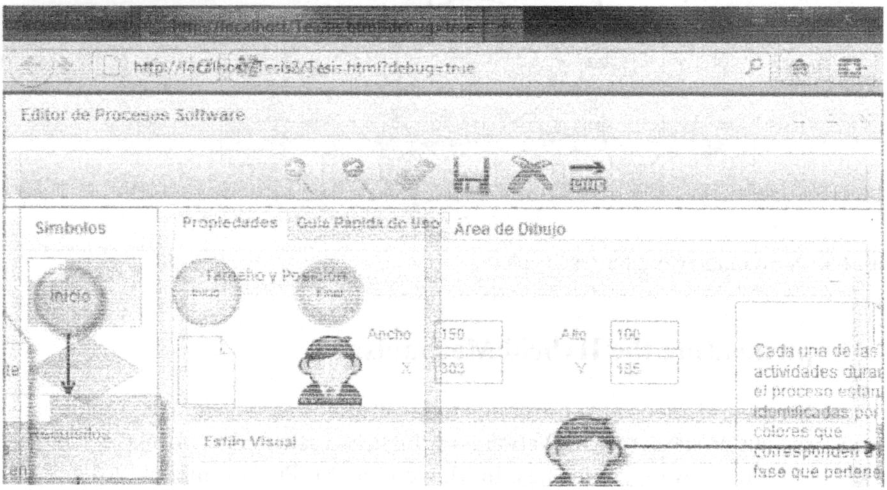

Fig. 3 SelfVation implementation for the modeling-based assessment

4.1 A RIA Implementation for the Modeling-Based Assessment

On the graduate program of Masters in Computer Science at the Technological University of the Mixtec Region (UTM) in Mexico, a research project was conducted to obtain data that contributed to the successful implementation of an assessment tool. The main tasks of the methodology for this project were:

1. Identification of, at least, one small software company which wanted to contribute with their knowledge of the reference model,
2. Using a tool to assess the organizations' actual situation, and
3. Collect information related to questionnaire-based and modeling-based assessments.

An assessment over Business Management Processes (GN), Specific Projects Management (APE) and Software Development and Maintenance (DMS) categories in Capability Level 2 was performed. This experimental assessment was performed two times: using the approach presented in [8] and the combined approach presented in this paper. Both assessments included a verification stage through verification questions and recollection of probatory documents. Table 1 summarises the assessment' characteristics. After, a workshop provided basic training in the MoProSoft model and in SPI; the assessed personnel were trained in using the designed RIA tool. Figure 4 shows the percentage obtained by each participant on the DMS assessment.

Table 1 Assessment' characteristics

Enterprise activity	Software development and maintenance
Type of assessment	Capability level
Reference model selected	MoProSoft
Assessed level	2
Assessed processes	GN, APE, DMS
Assessed personnel	1 top management member 3 project managers 3 programmers
Assessment conditions	Simultaneous assessment per groups, avoiding direct contact among them
Assessment duration	1 month

Figure 4 (a) summarises the results obtained using the traditional approach (two-phase questionnaire) implemented and adapted for the MoProSoft model. This approach is good; however it entirely depends on the commitment of personnel. The results show a coverage level of 49% for MoProSoft Capability Level 2. However, the verification phase demonstrated that this level did not correspond with the collected probatory documents for each assessed project. These results enable to us to establish a direct relation between the commitment level and answers' reliability. Different results were obtained with the proposed modeling-based approach (combining questionnaire and process modeling). Figure 4 (b) shows that the results obtained expose more accurate diagnostic criteria in accordance with the verification process performed after the assessment. Figure 5 shows that with the proposed approach through the RIA tool, the assessed personnel must firstly model their current software process in an individual way or in a collaborative way.

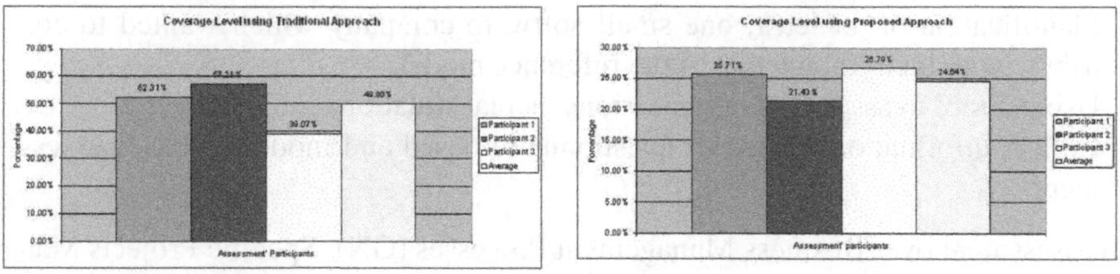

(a) (b)

Fig. 4 Assessment results

The modeling mechanism detects inconsistencies according to the selected reference model and qualifies this stage. The modeling stage selects only those questions which correspond with the current modeled process. The questionnaire mechanism assesses the execution level of the practice and qualifies the level of performance. The results show a coverage level of 24% for MoProSoft Capability Level 2. After implementing the hybrid mechanism we obtained information from the practices and effort performed by this small software company. More importantly, we collected information about the use of both assessment mechanisms: questionnaire-based and a mixed (questionnaire and modeling-based) approach. We assessed a total of 12 project managers in Capability Level 2 of the MoProSoft model. The experiment mixed all project managers (independent of their experience using MoProSoft) and assessed their knowledge in the model. We assumed that some of the project managers could be inaccurate about some questions; however we attempted to use the mixed application to improve the reliability of answers.

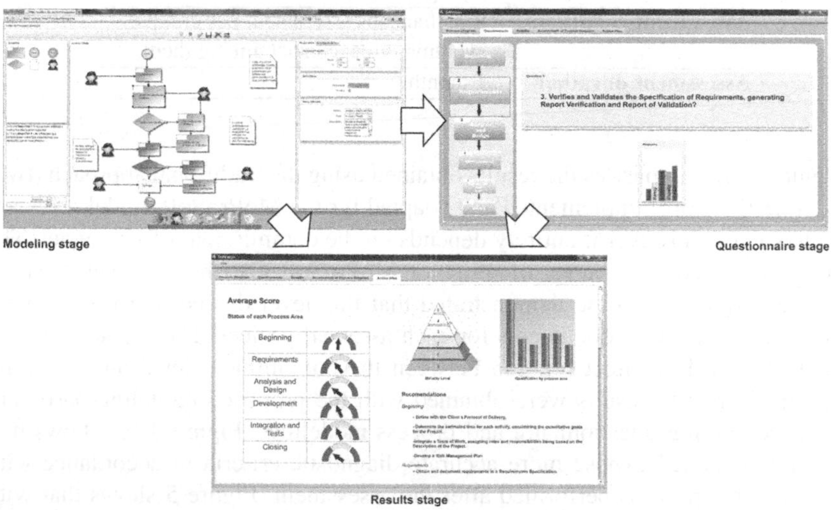

Fig. 5 Proposed approach used in process assessment

5 Conclusions

The implementation success when enterprises try to adopt an improvement initiative, depends on the commitment of the organization's top management. Small companies are not the exception, but this research enables them to define a modeling-based assessment mechanism for those evaluations where the benefits of an improvement process are not understood. To initiate the improvement program, involvement and experience of key personnel could be the key factors that contribute to strengthen the improvement initiative. But, the MoProSoft model is relatively new in Mexico and small software companies have no experience in SPI.

This research, therefore, has developed an instrument to start the implementation of SPI initiatives, and improve the current status of small software company practices using the MoProSoft as a reference model. Its purpose was to investigate the feasibility of the hybrid mechanism in small companies and to influence the direction of future research. One limitation of this study is the generalization of its findings based on the limited amount of data collected and analyzed relative to the number of small software companies. This suggests that this qualitative study will be augmented by quantitative studies to strengthen the data supporting the need and applicability of modeling-based assessment for the small Mexican company community.

References

1. Calvo-Manzano, J.A., Cuevas, G., San-Feliu, T., De-Amescua, A., Arcilla, M.M., Cerrada, J.A.: Lessons Learned in Software Process Improvement. The European Journal for the Informatics Professional 4, 26–29 (2003)
2. Cuevas, G., Serrano, A., Serrano, A.: Assessment of the Requirements Management Process using a Two-Stage Questionnaire. In: Proc. of the Fourth International Conference on Quality Software (QSIC 2004), pp. 110–116. IEEE Computer Society, Los Alamitos (2004)
3. Chrissis, M., Wemyss, G., Goldenson, D., Konrad, M., Smith, K., Svolou, A.: CMMI Interpretive Guidance Project: Preliminary Report (CMU/SEI-2003-SR-007). Software Engineering Institute, Carnegie Mellon University, Pittsburgh, PA (2003)
4. CMMI Product Team: CMMI for Systems Engineering, Software Engineering, Integrated Product and Process Development, and Supplier Sourcing (CMMISE/SW/IPPD/SS, V1.1). Continuous Representation. CMU/SEI-2002-TR-011, Software Engineering Institute, Carnegie Mellon University (2002)
5. CMMI Product Team.: CMMI for Development (CMMI-DEV, V1.2). CMU/SEI-2006 TR-008, Software Engineering Institute, Carnegie Mellon University (2006)
6. Dunaway, D.K., Masters, S.: CMM-based appraisal for internal process improvement (CBA-IPI). Method description. Technical Report. CMU/SEI-96-TR-007, Carnegie Mellon University, Software Engineering Institute (1996)
7. Flores, B., Astorga, M., Olguín, J., Andrade, M.C.: Experiences on the Implementation of MoProSoft and Assessment of Processes under the NMX-I-059/02-NYCE-2005 Standard in a Small Software Development Enterprise. In: Proc. of the 2008 Mexican International Conference on Computer Science, pp. 323–328. IEEE Computer Society Press, Los Alamitos (2008)

8. Fowler, P., Middlecoat, B., Yo, S.: Lessons Learned Collaborating on a Process for SPI at Xerox (CMU/SEI-99-TR-006, ADA373332), Software Engineering Institute, Carnegie Mellon University (1999)
9. Galliers, R.: Information Systems Research. Issues, Methods and Practical Guideline. Alfred Waller Ltd., Wiltshire, Chippenham. England (1992)
10. Garcia, I., Calvo-Manzano, J.A., Cuevas, G., San Feliu, T.: Determining Practice Achievement in Project Management Using a Two-Phase Questionnaire on Small and Medium Enterprises. In: Abrahamsson, P., Baddoo, N., Margaria, T., Messnarz, R. (eds.) EuroSPI 2007. LNCS, vol. 4764, pp. 46–58. Springer, Heidelberg (2007)
11. Humphrey, W.S.: Managing the Software Process. Addison Wesley, Reading (1989)
12. ISO/IEC 12207: 2008, Systems and software engineering – Software life cycle processes, Geneva (2008)
13. ISO/IEC TR 15504:1998(E): Information Technology – Software Process Assessments. Parts 1-9. International Organization for Standardization: Geneva (1998)
14. ISO/IEC 15504:2003/Cor.1:2004(E): Information Technology – Process Assessment. Parts 1-5. International Organization for Standardization: Geneva (2004)
15. Koch, N., Knapp, A., Zhang, G., Baumeister, H.: UML-Based Web Engineering: An Approach Based on Standards. In: Rossi, G., Pastor, O., Schwabe, D., Olsina, L. (eds.) Web Engineering: Modeling and Implementing Web Applications. HCI Series, pp. 157–191. Springer, Heidelberg (2007)
16. Kuvaja, P., Simila, J., Krzanik, L., Bicego, A., Koch, G., Saukonen, S.: Software Process Assessment and Improvement: The BOOTSTRAP Approach. Blackwell Publishers, Oxford (1994)
17. Linaje, M., Preciado, J.C., Sanchez-Figueroa, F.: Engineering Rich Internet Application User Interfaces over Legacy Web Models. Internet Computing Magazine IEEE 11, 53–59 (2007)
18. Marciniak, J.J., Sadauskas, T.: Use of Questionnaire-Based Appraisals in Process Improvement Programs. In: Proc. of the Second Annual Conference on the Acquisition of Software-Intensive Systems, Arlington, Virginia, USA, p. 22 (2003)
19. McCaffery, F., Taylor, P., Coleman, G.: Adept: A Unified Assessment Method for Small Software Companies. IEEE Software 24, 24–31 (2007)
20. NMX-I-059/01-NYCE-2005. Information Technology Software Process and Assessment Model to Software Development and Maintain Part 02: Processes requirements (MoProSoft). NMX-NYCE (2007)
21. NMX-I-059/04-NYCE-2005. Information Technology Software Process and Assessment Model to Software Development and Maintain Part 04: Guidelines for processes assessment (EvalProSoft). NMX-NYCE (2007)
22. Notkin, D.: Software, software engineering and software engineering research: some unconventional thoughts. Journal of Computer Science and Technology 24, 189–197 (2009)
23. Members of the Assessment Method Integrated Team: Standard CMMI® Appraisal Method for Process Improvement (SCAMPI), Version 1.1. CMU/SEI 2001-HB-001. Software Engineering Institute, Carnegie Mellon University (2001)
24. Preciado, J.C., Linaje, M., Morales-Chaparro, R., Sanchez-Figueroa, F., Zhang, G., Kroi, C., Koch, N.: Designing Rich Internet Applications Combining UWE and RUXMethod. In: Proc. of the Eighth International Conference on Web Engineering (ICWE 2008), pp. 148–154. IEEE Computer Society, Los Alamitos (2008)
25. Pries-Heje, J., Johansen, J., Christiansen, M., Korsaa, M.: The ImprovAbility Model. CrossTalk - The Journal of Defense Software Engineering 20, 23–28 (2007)

26. Santos, G., Montoni, M., Vasconcellos, J., Figuereido, S., Cabral, R., Cerdeiral, C., Katsurayama, A., Lupo, P., Zanetti, D., Rocha, A.: Implementing Software Process Improvement Initiatives in Small and Medium-Size Enterprises in Brazil. In: Proc. of the 6th International Conference on the Quality of Information and Communications Technology (QUATIC 2007), pp. 187–196. IEEE Computer Society, Los Alamitos (2007)
27. Weber, K.C., Araújo, E.E.R., da Rocha, A.R.C., Machado, C.A.F., Scalet, D., Salviano, C.F.: Brazilian Software Process Reference Model and Assessment Method. In: Yolum, p., Güngör, T., Gürgen, F., Özturan, C., et al. (eds.) ISCIS 2005. LNCS, vol. 3733, pp. 402–411. Springer, Heidelberg (2005)
28. Zubrow, D., Hayes, W., Siegel, J., Goldenson, D.: Maturity Questionnaire (CMU/SEI-94-SR-7). Software Engineering Institute, Carnegie Mellon University (1994)

Predicting 911 Calls Using Spatial Analysis

Dan Cramer, Albert Arthur Brown, and Gongzhu Hu

Abstract. A 911 call may be a result of an emergency medical need, fire attack, natural disaster, crime or an individual or group of persons needing some form of emergency assistance. Policy makers are normally faced with difficult decisions of providing resources to handle these emergencies, but due to lack of data and their inability to foresee the occurrences of these problems, they are caught by surprise. In this paper, we develop a model that will help policy makers anticipate the occurrences of emergencies. Spatial analysis methods such as hotspot analysis are used that can help policy makers distribute resources fairly by needs.

Keywords: Spatial analysis, stepwise regression, hotspot analysis, variable selection.

1 Introduction

A 911 call may be as a result of an emergency incident such as medical need or an occurrence of crime. Whatever the reason, it takes community resources to respond to these requests. How quickly the response will be depends on the resources and infrastructure availability. According to [11], 911 operations were constrained because of aged hardware and software, crude distributor, and lack of reliable data to manage operations of high volume of calls. Slow

Dan Cramer
Department of Geography, Central Michigan University, Mt. Pleasant, MI 48859, USA

Albert Arthur Brown · Gongzhu Hu
Department of Computer Science, Central Michigan University, Mt. Pleasant, MI 48859, USA
e-mail: hu1g@cmich.edu

R. Lee (Ed.): Software Eng. Research, Management & Appl. 2011, SCI 377, pp. 15–26.
springerlink.com

response rate to 911 calls can lead to death, destruction of property and high crime rate.

If policy makers are able to understand the factors which contribute high volume of 911 calls, they can implement proactive strategies to reduce the calls in the future. Understanding the factors can guide policy makers in distributing resources to community on basis of need.

This study is analyzes 911 emergency call data compiled from Portland Oregon metropolitan area. The study will seek to conduct a hotspot analysis to determine areas of high call volume, perform spatial analysis to understand factors that contribute these high call volumes and how they influence the number calls . Finally, best modeling approach when dealing with spatial data predict future calls will be developed based on the factors identified. We believe, this will help shape policy making and a guide to distributing resources to communities fairly.

2 Related Work

Some initiative taken to improve 911 call response times has been to address the problem of infrastructure and resource. According to [11], installation of enhanced 911 systems and comprehensive analysis of the communication division has been used to improve 911 operations at Washington D.C area. Although resources are necessary, it is not a proactive way to reduce the high volume of 911 calls to an area and does not try to understand the problem in other to address it. Proactive approach will ensure early identification of hotspots and when addressed will lead less resource being used. Hotspot analysis is to study about concentrations of incidents or events within a limited geographical area over time [8, 13].

Being proactive means identifying a strategy to anticipate the call volume in a given location. Regression analysis can be used to investigate the relation between geographic variable as well predict future volume of call. For example, a multi-linear regression model [10] was proposed to predict the number of 911 calls for a period of interest in the San Francisco area. The model is of the form:

$$y = a_0 + a_1x_1 + a_2x_2 + ... + a_8x_8 \tag{1}$$

where y is the number of calls for the time period of interest, and x_1 through x_8 are the number of calls for the previous 8 times. In this model, the predictions are used to detect possible emergency events by sounding an alarm when the number of calls exceeds the predicted number of calls. Both linear regression and geographically weighted regressions can be used. However, when location has to be accounted for, then linear regression has a limitation because it does not account for the variability of a spatial context [2]. Several approaches have been used to account for location in spatial analysis of relationship between variables. For example, Cassetti [1] proposed the expansion method where coefficients in regression model are expressed as explicit functions of the spatial locations of the cases:

$$y_i = \sum X_{ij}\beta_j(p_i) + \epsilon_i \tag{2}$$

where p_i is the geographical location of the i-th case. The $\beta_j(p_i)$'s would contain the coefficients to be estimated. These expressions would be substituted into the original model and the resulting model will be expanded. Another approach proposed by Brunsdont [2] is to include dummy variables for broad classes of spatial location, e.g. to indicate which county a case is situated or whether the location could be classified as rural or urban. This approach had a limitation because it does not allow other coefficients in the model to vary spatially. Gorr and Olligschlaegar [7] also proposed a method using spatial adaptive filtering to model spatial variations in coefficients, but this method does not have any means of statistically testing the validity of this assumption of variation in parameters. According to [2], the approaches mentioned above, relied on spatial hierarchy and no research had been done to investigate the sensitivity of these models to changes in the hierarchical grouping of the zones. One method which builds on the method proposed by Cassetti [1], is the Geographically Weighted Regression (GWR) [5]. It attempts to overcome the limitation of the Cassetti approach by providing a non parametric estimate of $\beta_j(pi)$ [2]. Weighted Regression analysis can model, examine and explore spatial relations to better understand the factors behind spatial patterns.

Other approaches have been proposed for predicting the future call volumes, such as 911 call stream analysis [9] and time series [12].

3 Data

The data we used in our analysis is the 911 calls from 1/1/1998 – 5/31/1998 from Portland, Oregon Metro Area, downloaded from [6], the latest data set we could find that fit our need. It consists of 2000 calls. The data was already

Table 1 Variable Description

Variable Name	Description
Renters	Number of people renting in each tract
Businesses	Number of businesses in each tract
Jobs	Number of jobs in each tract
College Graduates	Number of college graduates (bachelors or higher) in each tract
NotILF	Not In Labor Force
Dist2UrbCenter	Distance to urban center

gecoded with responses addresses and the unit data collection was census tract. The most important variables of the dataset are shown in Table 1.

4 Methodology

The results of any research experiment are only as good as the methods and processes that were used to create them. This section discusses the methods of how spatial analysis tools and functions were carried out in order to analyze 911 call hotspots and the regression analysis that followed.

The primary goal of this analysis is to first determine hotspots of 911 calls. The next part of the analysis is to determine what factors have an influence on the number of 911 calls. And finally the third part of the analysis is to determine geographically how these factors influence the number of 911 calls.

4.1 Data Preprocessing

As with any dataset, there is almost certainly a great deal of cleaning and preprocessing before any real analysis can be done. The preprocessing stage can often take the majority of the time spent on a data analysis project. Fortunately for the sake of the analysis this part didn't consume as much time as it could have due to the data being fairly clean to begin with. After downloading the data one of the first things one should do is examine the data and see what is there. The data file contained a ESRI shapefile (point layer) of 911 calls in the Portland Oregon Metropolitan Area. In all, there were over 2000 911 calls in the dataset taken from 11/1998 to 5/31/1998. By examining the 911 call dataset and reading the literature associated with the data, the 911 calls were already geocoded. This was an immediate resource saver because the geocoding process is a very lengthy and time consuming process. The other file associated with the 911 call dataset was a polygon file of Census Tracts. The data was from the 2000 Census. This file also contained a lot of data that wouldn't necessarily be time consuming to obtain, but it was already processed with many socioeconomic variables.

Another piece of data that was with the data set was a point file of three 911 Response Dispatch Centers. These centers are where the 911 calls were taken from. For the analysis, the only important part of this file was the primary location of the response center. There were also some miscellaneous files such as road network, but that was not of importance in the primary analysis. Before any of the analysis was to take place, it was necessary to get the total number of 911 calls into each census tract. This was done by an operation in ArcMap version 9.3 called a spatial join. A spatial join creates a table join in which fields from one layer's attribute table are appended to another layer's attribute table based on the relative locations of the features in the two layers [3]. The spatial join takes calls that intersect each tract and then calculates a count.

4.2 Variable Selection Using Stepwise Regression

The next step in the analysis was variable selection. For this analysis, a stepwise regression was done in Minitab version 16. Since there were a relatively large number of variables, picking and choosing all of the different combinations of variables would not only be a waste of time, but could possibly yield a misleading model. Stepwise regression determines significant independent variables in an automated process. In a stepwise regression, independent variables are brought into the model and taken out by the model based on certain criteria. Minitab defaults to an alpha of 0.15. While this is somewhat conservative as far as significance goes, it allows model selection to at least be done, and then interpreted by the analyst.

The stepwise regression result showed that renters, distance to urban center, businesses, jobs, not in labor force, and college graduates were the independent variables in the model. It also showed that this model had a standard error of 10, a R Squared of 88.4%, and a CP of 5. Of course just because a statistical software package picks out a model it doesnt mean that it is a good model, but it gives a starting point.

Now that there was an idea of what variables were of importance, diagnostics and remedial measures, if any, needed to be looked at. The first step in analyzing the model and its variables was to do a correlation analysis.

This was done in Minitab just like the stepwise regression. When looking at the correlation matrix from the output it was important to see if any variables were highly correlated with each other. It was determined that no variables were so highly correlated with each other that it would cause problems in the model.

After the correlation analysis, scatter plots and histograms of each variable were examined. This was done in ArcMap through a scatter plot matrix program. The program allowed checking different variables against each other to determine if there was any high correlation. While some of the variables exhibiting a noticeable amount of correlation, nothing was so correlated that it warranted a transformation or removal.

As a final preprocessing/data visualization process, we created maps of each variable in the model. This was also done in ArcMap. This part is extremely beneficial and is one of the advantages of using GIS in any project. Using GIS allows for an overall interpretation of how variables are distributed.

4.3 Hotspot Analysis

The data has been preprocessed, examined, cleaned, and is now ready for analysis. As mentioned before the scope of this analysis was to first do a hotspot analysis, then do a linear regression, and then a geographically weighted regression.The hotspot analysis was run in ArcMap. The technical name of the hotspot analysis is the Getis-Ord Gi* Hot Spot Analysis [4]. The Getis-Ord Hotspot Analysis tool works by, when given a set of weighted data points, the Getis-Ord Gi* statistic identifies those clusters of points with values higher in magnitude than you might expect to find by random chance [3].

The output of the Gi function is a z–score for each feature. The z–score represents the statistical significance of clustering for a specified distance. High z–score for a feature indicates its neighbors have high attribute values, and vice versa. The higher (or lower) the z–score, the stronger the association. A z–score near zero indicates no apparent concentration. After running the hotspot analysis and determine areas of high or low calls, it was then appropriate to determine what factors had an influence on the number of calls.

After running the hotspot analysis and determining areas of high or low calls, it was then appropriate to determine what factors had an influence on the number of calls. The next major part of the analysis was to then determine what variables had a significant influence on the number of 911 calls. The statistical method for attempting to answer this question is by performing a linear regression. Since we had already done variable selection with the stepwise regression model, this part of the analysis was pretty straightforward. We ran the model with the variables that were earlier decided to be significant using Minitab and then also ran the model in ArcMap. The purpose of running the model twice was because Minitab allows for more options in outputs and diagnostics. After running the model in Minitab and closely observing the diagnostics, the model was run in ArcMap to give a spatial interpretation of the model.

4.4 Geographic Weighted Regression

The next phase of the analysis was to use a Geographically Weighted Regression(GWR) Model to determine how each variable had an influence over the different spatial extents. A GWR is different from a least squares regression in that it is a local model where a least squares regression is a global model. For example, renters could have a higher influence in one area and a much lower influence in another area. A GWR will be able to determine this. The GWR was run in ArcMap with the same variables as the least squares regression model using inverse distance as the distance metric. The equation for a GWR is:

$$y_i = \beta_0(u_i, v_i) + \sum \beta_k(u_i, v_i) + \epsilon_i \tag{3}$$

After both models were run in ArcMap, a check for spatial autocorrelation was done to determine if there was any autocorrelation in the residuals. This diagnostic uses the Local Moran's I function.

5 Result

5.1 HotSpot Analysis

Fig. 1 shows the result of the hotspot analysis to identify area of high call volumes and low call volumes. The test statistic in hot spot analysis is the

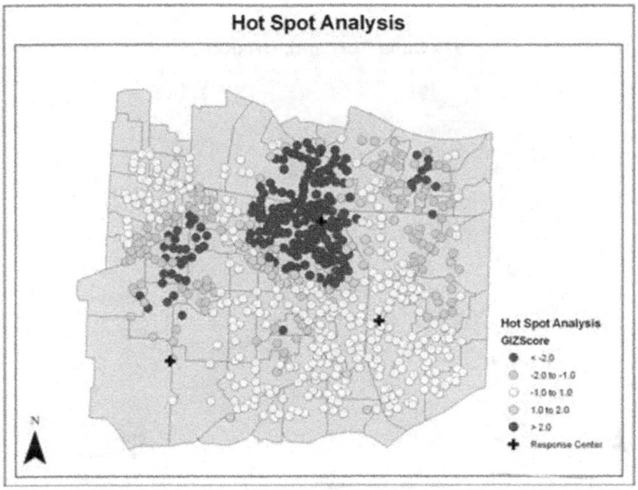

Fig. 1 Hotspot Analysis

Getis-Ord Gi* statistic, which identifies those clusters of points with values higher in magnitude than you might expect to find by random chance. z–scores less than –1.96 and greater than 1.96 were significant at 95% confidece interval. Areas with higher z–score are area with high call volume and areas with low z–score have relatively lesser call volume and represent cold spot.

5.2 Least Square Regression

The regression model used to determine what factors had an influence on 911 calls is:

$$\begin{aligned} Calls = {} & 16.0 + 0.0211\, Renters \\ & -0.00149\, Dst2UrbCen + 0.0360\, Businesses \\ & +0.00268\, Jobs + 0.0625\, NotInLF \\ & -0.0291\, CollGrads. \end{aligned} \tag{4}$$

The map of the residuals in Fig. 2 shows a visual output of the regression. Areas in red are under predications, and areas in blue are over predictions. This is similar to a fitted regression line output. The regression outputs in ArcMap show the same statistics as Minitab, and even introduce a few more, but this analysis only focused on model selection and validity to answer the question of what variables influence 911 calls.The R-squared of the model was 88.4%, which means that 88.4% of the model can explained by the predictors in it. A high R-sqaured of the model is an indicator of a good model. The model also had an AIC 655 , and a standard error of 10.8. The check for

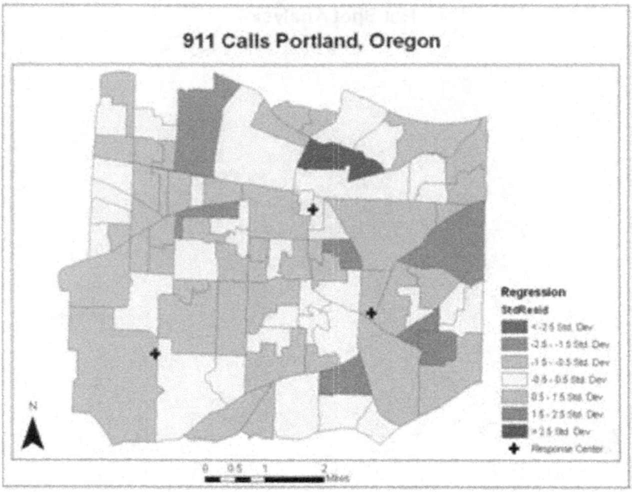

Fig. 2 Least Square Regression Output

autocorrelation on the residuals showed a random pattern, which was ideal and somewhat expected based on the normality of the residuals from the Minitab output, shown in Fig. 3.

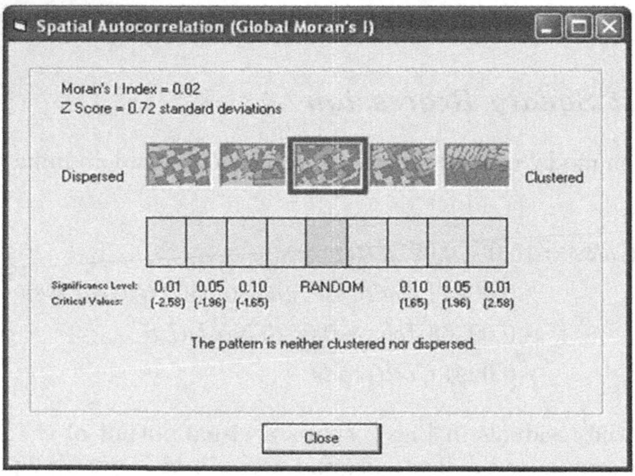

Fig. 3 Autocorrellation Output

5.3 Geographically Weighted Regression

In order to determine the spatial variability in the model a GWR had to be run. This was done exclusively in ArcMap. The outputs of the GWR on the surface look similar to the least squares regression model, especially from a map point of view, but they are fundamentally very different. A least squares regression is global model. This type of model takes the entire area into consideration. The GWR is a local model; therefore it takes into consideration each unique spatial unit, which in this analysis were census tracts. The GWR is an ideal model when working with spatial datasets because of the flexibility that it provides versus a least squares model. Usually, although not all of the time a GWR will be a better model. The GWR model had a R square value of 91.04%, AIC of 658.19, and a standard error of 9.55. The residuals of the GWR model were also checked using the local Moran's I function in ArcMap. The residuals were random, just like the least squares model.

5.4 Model Comparison

One of the main goals of this analysis was to determine if one type of regression model was more optimal than another. The goal was to determine what model was better when working with spatial data. In addition to the GWR model, we slaso run tests to calculate some statistics measures for ordinary least squares (OLS) regression. The comparison of the two models is given in Table 2. Based on the both model outputs, it appeared as if the GWR model was a better model with this data.

Table 2 Comparison of ordinary least squares model and GWR

Model	OLS	GWR
R Square	88.40%	91.04%
Adjusted R Square	87.50%	88.80%
AIC	655.83	658.19
Standard Error	10.80	9.55

The GWR model had better criterion for the model with the exception of AIC which was just slightly lower in the least squares model. This wasn't big enough of a factor to make the decision that the GWR wasn't as strong of a model compared to the least squares model, especially since the other criterion were higher with the GWR model.

6 Discussion

While this analysis had three major parts to it, there was one overall goal and that was to predict and better analyze 911 calls through spatial analysis. The hotspot analysis was done to determine areas of high and low concentrations of calls, and the regression models were used to determine which factors.

The hotspot analysis showed where concentrations of 911 calls were. This is particularly important to many people because if they can determine which areas have high or lower calls, then they can better use resources towards responding to emergencies. For example, the response center in the southwest region of the map has statistically significant cold spots around it, while the call center in the middle of the map has a large amount of statistically significant hot spots. It is tough to tell with just 5 months of data, but perhaps more resources could be sent to the call center with high call volume and less towards the one with low call volume. This type of policy decision is out of reach for our analysis, but that could be a potential implication of an analysis like this.

The variables used in the regression analysis were chosen by their statistical significance, but weren't explained by the data processing software packages. The variables used in the regression models also need to be interpreted to give context to the analysis. The variables used once again were, jobs, renters, distance to urban center, not in labor force, businesses, and college graduates. The variables, jobs, renters, not in labor force, and businesses were positive, and the variables college graduates and were negative. This is interpreted as when call increase by one, the independent variables increase or decrease by their respective coefficients.

In the pre-processing stage of the analysis, we created maps of each variable. By viewing this data, we could tell that areas in the center of the map were the major populated areas of Portland, thus it makes sense for the hotspots of calls to be there. Therefore if there is a high population area, then there is going to be more businesses, more jobs, and more rentable properties and so on. As an additional option in the GWR model, coefficient rasters could be generated to show the spatial distribution of the variables in the model. These raster outputs show a range of all the coefficients in the model so it is easy to see where they are distributed spatially. For example the variable renters were more distributed in the south west portion of the Portland area as shown in Fig. 4 below.

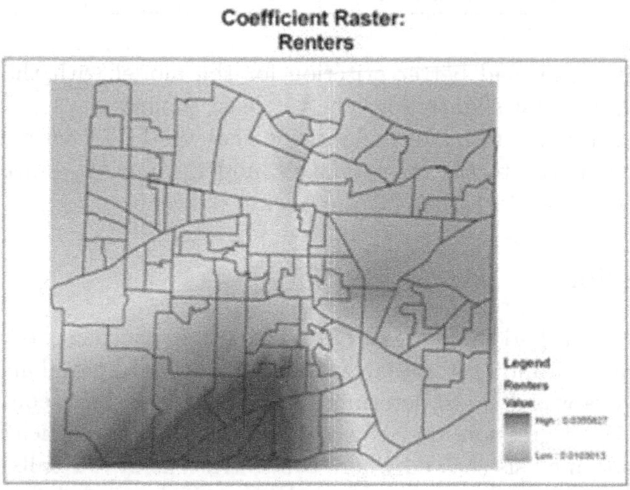

Fig. 4 Distribution of Renters

The advantage to using the GWR model over the least squares model is that the GWR model takes each spatial component into consideration when building the model.

7 Conclusion and Future Work

This paper provided some insights into how spatial analysis is applied to a dataset. This analysis attempted to determine hot spots of 911 calls and then interpret what factors have an influence in them. The major goals of this work were successively accomplished for the sake of data analysis as far as results and outputs go. However, we didn't attempt to make any policy decisions based on our results. There is a potential for determining what actions could follow up the results that were found in this analysis.

Spatial analysis provides for powerful data analysis. The ability to combine statistical analysis as well as strong visualizations for it certainly makes an impressive analysis. Many times in a data analysis project, numbers and figures get overwhelming not only for the analyst, but also for the audience. The ability to provide access to the data analysis through maps and charts makes the data easier to understand.

A major limitation on this project was the spatial and temporal aspects of the data. The data was only for a portion of the Portland, Oregon area, and the call data was from 5 months of collection. More data over a larger area could possibly lead to different results. Perhaps the data in the Portland area is different from other cities. Also because the data was relatively small, applying the model across other areas may not lead to a valid and acceptable model.

References

1. Cassetti, E.: The expansion method. Journal of Information Science, 432–449 (1972)
2. Chris, B., Martin, C., Stewart, F.: Geographically weigthed regresion - modelling spatial non stationarity. The Statistician, 431–443 (1998)
3. ESRI: ArcGIS Desktop Help 9.3 (2009), `http://webhelp.esri.com/arcgisdesktop/9.3`
4. ESRI Developer network: Hot Spot Analysis (Getis-Ord Gi*) (spatial statistics), `http://edndoc.esri.com/arcobjects/9.2`
5. Fotheringham, A.S., Brunsdon, C., Charlton, M.: Geographically Weighted Regression: The Analysis of Spatially Varying Relationships (2002)
6. GIS Blog: GIS Data, `http://www.gisiana.com`
7. Gorr, W.L., Olligschlaeger, A.M.: Weighted spatial adaptive filtering: Monte Carlo studies and application to illicit drug market modeling. Geography Analysis, 67–87 (1994)
8. Grubesic, T.H., Murray, A.T.: Detecting hot spots using cluster analysis and gis. In: Proceedings from the Fifth Annual International Crime Mapping Research Conference (2001)

9. Hodgkiss, W., Baru, C., Fountain, T.R., Reich, D., Warner, K., Glasscock, M.: Spatiotemporal analysis of 9-1-1 call stream data. In: Proceedings of the 2005 National Conference on Digital Government Research, pp. 293–294, Digital Government Society of North America (2005)
10. Jasso, H., Fountai, T., Baru, C., Hodgkiss, W., Reich, D., Warner, K.: Prediction of 9-1-1 call volumes for emergency event detection. In: Proceedings of the 8th Annual International Conference on Digital Government Research: Bridging Disciplines & Domains, pp. 148–154. ACM, New York (2007)
11. Kuhn, P., Hoey, T.P.: Improving 911 police operation, pp. 125–130. National Productivity, Washington, D.C (1987)
12. Tandberg, D., Easom, L.J., Qualls, C.: Time series forecasts of poison center call volume. Clinical Toxicology 33(1), 11–18 (1995)
13. Zeng, D., Chang, W., Chen, H.: A comparative study of spatio-temporal hotspot analysis techniques in security informatics. In: Proceedings of 7th International IEEE Conference on Intelligent Transportation Systems, pp. 106–111. IEEE, Los Alamitos (2004)

A Market-Driven Product Line Scoping

Kwangchun Lee and Dan Hyung Lee

Abstract. As markets fragmented into a plethora of submarkets, coping with various customers demands is becoming time consuming and expensive. Recently software product line (SPL) approach has shown many benefits as opposed to single product development approach in terms of quality, time-to-market, and cost. However, quantitative market-driven scoping method from existing or future product portfolio and relating customers to products has not been explored. Even though setting the proper scope of the product line is the first step to establishing initiatives in software product line, a market-driven scoping based on engineering principles has not been fully exploited. In this paper, we propose a market-driven quantitative scoping method. This method incorporates customers' needs, product family structure and market strategies into scoping such that this ensures that SPL derivatives penetrate market grids.

Keywords: software product line engineering; scoping; share of wallet; life time value; leveraging strategy.

1 Introduction

Recently software products are complex in nature and must provide extensive functions to users in order to make them useful. There have been numerous approaches to cope with software quality challenges such as 1960's subroutines, 1970's modules, 1980's object, and 1990's components. However, providing these functions always has shown diminishing returns. One of ways circumventing these technological drawbacks is to utilize software product lines.

Instead of building a single software product, product family and line concept can increase software development productivity, reduce lead time and improve quality. A software product line is a set of software-intensive systems sharing a common, managed set of features that satisfy the specific needs of a particular market segment or mission and that are developed from a common set of core assets in a prescribed way [4].

Kwangchun Lee · Dan Hyung Lee
Department of Computer Science, Korea Advanced Institute of Science and Technology, 373-1 Guseong-dong, Yuseong-gu, Daejon 305-701, Republic of Korea
e-mail: {statkclee,danlee}@kaist.ac.kr

R. Lee (Ed.): Software Eng. Research, Management & Appl. 2011, SCI 377, pp. 27–46.
springerlink.com

The opportunity for successful projects is at a project's start in the requirements [42] and a Standish Group Report supported this finding based on two of the top reasons for successful project criteria: user involvement and requirements scoping [38]. Early user involvement increases project success probability through requirements quality improvement and requirements scoping.

Given the time and resources available determining right products scope should consider three main activities: establishing relative priorities for requirements, estimate the resources needed to satisfy each requirement, and selecting a subset of requirements that optimizes the probability of the product's success in its intended market [6]. If the scope is chosen too large, this will result in a waste of investment on assets whereas if the scope is chosen too narrow, probabilistic reuse opportunities will be missed [34]. Systematic and practical guides to product line scoping have been described under the PuLSE framework [8,16,34].

The introduction of marketing concept and mass customization in software has changed the way to develop software products. Design issues from the marketing perspective have been explored and key design drivers that are tightly coupled with the marketing strategy have been presented [17]. In addition, how QFD (Quality Function Deployment) for product portfolio planning to identify customers and needs, derive a product portfolio systematically, and to derive common and variable product functions has been explored [10]. Recently, two different product lines, called Engineered Product Line and Marketed Product Line, are defined and gains from aligning scoping and product management have been described [11]. Product line balancing for allocating features to product lines has been used to facilitate effective deployment of complex product lines [41].

However, quantitative market-driven scoping methods from existing or future product portfolio and how to relate customers to products have not been explored. Since software product features are multi-dimensional and non numeric in nature, features has not been fully exploited as means of identifying product structure and relating customer segments to customized products. Even though setting the proper scope of the product line is the first step to establishing initiatives in software product line, a market-driven scoping based on engineering principles has not been fully exploited.

In this paper, we propose a market-driven quantitative scoping method. The proposed method is based on modern market analysis, mass customization strategies and advanced statistical methods such that incorporation of customers' needs, product family structure and market leveraging strategies into product line scoping enables us to develop customized products for target customer segments.

The structure of this paper is as follows. In section 2, we review some related works concerning mass customization and product line strategies. A framework for product line scoping to determine appropriate scope of product line is described in section 3. Section 4 demonstrates determination of commonality and variability for a product line of text editors. A market-driven quantitative scoping framework reflecting product line leveraging strategies and offering customized products to customer segments is described in section 5. Section 6 mentions future direction. Finally, section 7 outlines the conclusion.

2 Related Works

Mass customization aims to take advantages of mass production and customization through maximization of reuse. Since the mass customization (MC) was coined by Stan Davis [7], the concept has been widely accepted. Mass production pursues efficiency through process stability and control, whereas mass customization seeks to obtain variety and customization through flexibility and speediness. The benefits from mass customization include reduction of development time/cost and improvement of quality.

A comprehensive review of the product family engineering has been done by [15,24,37]. In the heart of mass customization there are commonality and modularity concepts. Commonality aims to decompose product family into unique and common parts. Common parts play a pivotal role as platform in product family. Products are derived from adding variable parts such that reuse maximization strategy can be institutionalized. Even though modularization shares objectives of commonality concept, it seeks to enjoy benefits of delayed postponement through latter binding and standardization. Components are designed and developed in accord with standard templates and interface definitions such that when purchasing order arrives, components are assembled and ready to sell.

There are two common approaches for product family development: top-down and bottom-up. The top-down approach usually takes high-risk/high-return strategy in that a platform should be constructed ahead of product realizations but the global optimum can be achieved. The bottom-up strategy redesigns existing products to standardize components such that introduction risk of platform can be reduced but there is a risk of local optimum.

The product line leveraging strategy and product positioning from mass customization can have benefits through advanced statistical methods. The product positioning and design problem from theoretical viewpoints has been examined by [19]. Wheelwright and Sasser [43] proposed leveraging strategy based on product family map and Meyer and Tertzakian [32] suggested methods to measure the performance of research and development for new product development in the context of evolving product families. Product family leveraging strategy through market segmentation grid has been proposed by [31].

3 A Framework for Product Line Scoping

Product features are identified and classified in terms of capabilities, domain technologies, implementation techniques, and operating environments [27]. Product features can be used as a communication facility such that the gap between customers and producers can be filled with clear understanding without energy exhaustion. Product features are usually represented as numeric, ordinal and nominal formats in the order of information amount. Numeric data representation makes it possible to do advanced analysis without loss of information, but collecting numeric data and designing numeric data generation process are time consuming and expensive. Ordinal and nominal data have broad representation of products and customer needs, but analyzing categorical data demand experts and high computing powers. However the advanced of desktop computing and internet have empowered users to deal with categorical data and extract information.

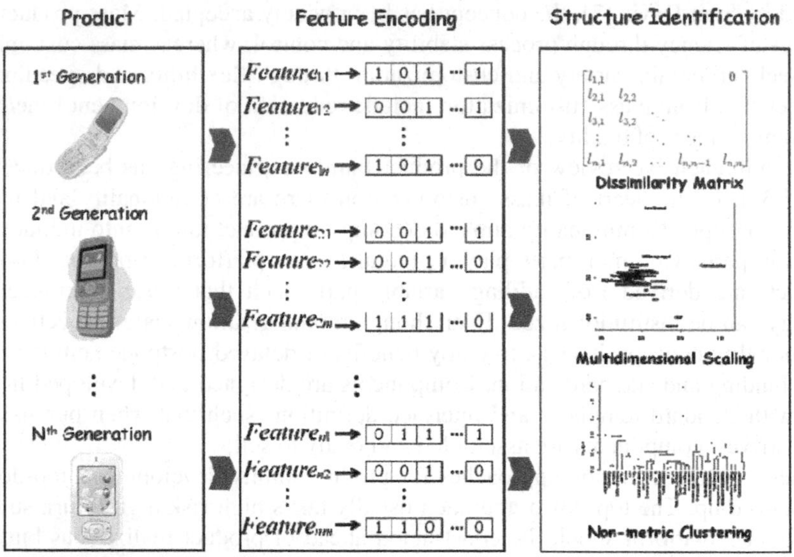

Fig. 1 A framework for product line scoping

A new method to reveal structure of product portfolios for commonality and variability identification through multidimensional scaling and non-metric clustering has been shown in Fig. 1. Products features are extracted and encoded in a binary format. Even though loss of information can occur when encoding products features in binary form, introducing domain knowledge in a binary encoding can reduce overall information loss. Dissimilarity matrix which is an input for further statistical analysis can be obtained from the binary encoding data and similarity distance measures. Revealing structure of product portfolios can be attributed to advanced statistical analysis such as multidimensional scaling and non-metric clustering.

3.1 Binary Features Encoding and Dissimilarity Matrix

Even though the binary representation of product features has many benefits, extracting information from binary data needs additional steps to reach goals. There are several ways to represent features into binary format. The goal should be clearly stated and data cleaning and preprocessing should be carried out cautiously so as to minimize information loss from the binary encoding. An example of binary encoding from software family of text editors has been shown in Fig. 2.

	Basic Feature			
	Spell checking	Regex-based find & replace	Encoding conversion	Newline conversion
Acme	No	Yes	No	No
Alphatk	Yes	Yes	Yes	Yes
Aquamacs	Yes	Yes	Yes	Yes
BBEdit	Yes	Yes	Yes	Yes
Bluefish	Yes	Yes	Yes	?
ConTEXT	No	Yes	Partial [15]	Yes
Crimson Editor	Yes	Yes	Yes	Yes
Diakonos	No	Yes	No	No
e	Yes[14]	Yes	Yes	Yes
ed	No	Yes	No	No
EditPlus	Yes	Yes	Yes	Yes
EmEditor Pro	No	Yes	Yes	Yes

	Basic Feature			
	Spell checking	Regex-based find & replace	Encoding conversion	Newline conversion
Acme	0	1	0	0
Alphatk	1	1	1	1
Aquamacs	1	1	1	1
BBEdit	1	1	1	1
Bluefish	1	1	1	0
ConTEXT	0	1	0	1
Crimson Editor	1	1	1	1
Diakonos	0	1	0	0
e	1	1	1	1
ed	0	1	0	0
EditPlus	1	1	1	1
EmEditor Pro	0	1	1	1

Fig. 2 Binary encoding illustration with text editors data

As a feature similarity measure binary distance measures such as Jaccard distance [13] can be used to transform the binary encoded data into dissimilarity matrix. Given two features F_1 and F_2 Jaccard distance can be defined as follows;

$$Jaccard\ Distance\ (F_1, F_2) = 1 - \frac{|F_1 \cap F_2|}{|F_1 \cup F_2|}$$

Instead of comparing two features, multiple features comparison will come up with dissimilarity matrix shown in Fig. 3.

	Single document interface	Single document window	Overlappable windows	Tabbed document interface	…	DOS (CR/LF)	Unix (LF)	Mac (CR)
Single document interface	0.000							
Single document window splitting	0.524	0.000						
Overlappable windows	0.698	0.452	0.000					
Tabbed document interface	0.500	0.382	0.419	0.000				
Window splitting	0.585	0.148	0.479	0.455				
Spell checking	0.395	0.351	0.472	0.368				
Regex-based find & replace	0.341	0.333	0.525	0.350				
Encoding conversion	0.455	0.421	0.412	0.351				
Newline conversion	0.413	0.375	0.525	0.390				
Syntax highlighting	0.341	0.375	0.487	0.308				
Multiple undo/redo	0.326	0.357	0.500	0.333				
Rectangular block selection	0.535	0.333	0.469	0.400				
⋮	⋮				…			
DOS (CR/LF)	0.391	0.390	0.500	0.405		0.000		
Unix (LF)	0.326	0.357	0.535	0.409		0.116	0.000	
Mac (CR)	0.457	0.385	0.459	0.359		0.077	0.143	0.000

Fig. 3 Dissimilarity matrix of text editor data based on jaccard distance

3.2 Revealing Structure through Multidimensional Scaling (MDS) and Non-Metric Clustering

Multidimensional scaling (MDS) [1, 25, 26, 30, 35, 36, 39, 40] has been proposed to reveal the discovery and representation of hidden feature structure underlying dissimilarity matrices. MDS aims to transform a set of high-dimensional data to lower dimensions in order to determine the dimensionality necessary to account for dissimilarities and to obtain coordinates within this space.

The Minkowski distance metric specifies the distance between two objects in multidimensional space:

$$d_{ij} = \left[\sum_{k=1}^{n} \left| x_{ik} - x_{jk} \right|^{r} \right]^{1/r}$$

where n is the number of dimensions and x_{ik} is the value of dimension k for object i.

There are two types of multidimensional scaling: metric and non-metric scaling. The metric or classical scaling assumes that the dissimilarities satisfy the metric inequality whereas the non-metric scaling only assumes that the data measured at the ordinal level and keep the rank order of the dissimilarities.

$$Stress = \sqrt{\frac{\sum_{i=1}^{n-1} \sum_{j=i+1}^{n} \left(d_{ij} - d_{ij}^{*} \right)^2}{\sum_{i=1}^{n-1} \sum_{j=i+1}^{n} d_{ij}^{2}}}$$

Kruskal's stress (Standardized REsidual Sum of Squares) measures 'goodness-of-fit' between the distances and the observed dissimilarities. Since there is no analytic solution in the non-metric MDS, an iterative approach should be adopted. While smaller stress means a better fit, higher dimensionality prevent revealing hidden structure in the dissimilarity data such that the principles of parsimony, or Occam's razor, is applied here.

The true dimensionality will be revealed by the rate of decline of stress as dimensionality increases. The scree plot where stress is plotted against the dimensionality is often used to select the dimensionality from the elbow in the plot because the stress or other lack-of-fit measures can be reduced substantially.

The quality of the fit can be visually evaluated through Shepard diagram. The Shepard diagram is a scatterplot of fitted distances on the vertical axis against the original dissimilarity on the horizontal axis. If the points in the Shepard diagram are close to the diagonal, it indicates a relatively good representation given dimensionality.

As opposed to monothetic analysis [18] which considers only a well selected feature at a time sequentially, non-metric MDS tries to preserve the structure of the original dissimilarities as much as possible and the representation of dissimilarity data from dimensionality selection and quality evaluation reveals hidden structures.

Clustering seeks to reduce development cost by avoiding duplicated efforts. There are two types of clustering [5,9,44]; hierarchical clustering and k-means clustering. Under hierarchical clustering each point is a separate cluster such that the two closest points are merged through single linkage, average linkage, and complete linkage until the desired number of clusters is obtained. K-means clustering [28] starts with randomly selected k initial centers and minimizes the objective functions.

$$O = \sum_{j=1}^{k} \sum_{i=1}^{n} \left(x_i^{(j)} - m_j \right)^2$$

where m_j is the cluster center and the Euclidean distance is assumed. Practitioners or researchers applying clustering techniques in practice deals with many issues associated with number of clusters, clustering methods, outliers, weights, and validation.

As opposed to non-metric MDS which tries to preserve the structure of the original dissimilarities as much as possible and Euclidean distance measures, the categorical data clustering provides different mechanism to reveal hidden structures. Particularly huge amount of non-metric type data is coming from software product features such that clustering with similarity/dissimilarity measures contributes to reveal hidden structures of product portfolios.

4 A Case Study: Commonality and Variability for a Product Line of Text Editors

The proposed method starts from the feature extraction of existing or concept products. The features of products are dissected and recorded in the binary format. The binary features of the products are transformed into matrix formats for feature similarity grouping and feature clustering.

The dissimilarity matrix is an input to MDS analysis and any statistical software packages including ALSCAL algorithm [39] can process the dissimilarity data. In order to find best representation of structure, iterative data analysis equipped with built-in Sheppard diagram, goodness-of-fit statistics, scree plot, and computational convergence is essential.

As an illustration, text editors from commercial and open source software have been collected. The text editor data include broad-range features such as platforms, languages, document interface, basic features, programming features, extra features, key bindings, protocol support, encoding support, newline and support. Since the majority of features are represented in the non-numeric format, classical and traditional analysis methods cannot be applied.

Due to the characteristics of non-numeric data and high dimensionality, it is necessary to apply non-metric dimension reduction techniques. The first step is to recode text editor features into binary which typically leads to many zeros and ones. Once features of text editors have been encoded, suitable metrics for similarity of features from text editors should be defined. Even though there are many distance metrics, Jaccard distance is commonly used for distance measures encoded in a 0-1format.

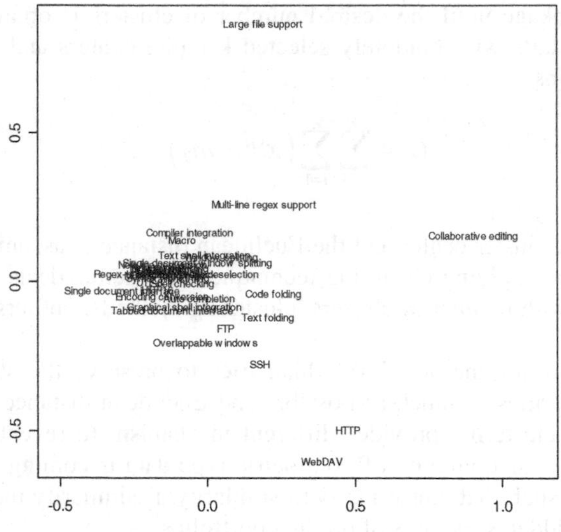

Fig. 4 Non-metric MDS analysis of text editors

The calculation of Jaccard distances lead to dissimilarity matrix. Once the dissimilarity matrix has been obtained, non-metric MDS analysis reveals hidden relationship among text editor features. Due to high dimensionality of text editor family, low dimensional projection is needed for visual inspection and explanation. The Fig. 4 shows that variability features include 'large file support', 'collaborative editing', 'HTTP protocol', and 'WebDAV'. In addition to optional features, pseudo core features include 'SSH', 'text folding', 'code folding', 'FTP', 'overlappable window', and 'multiline regex support'.

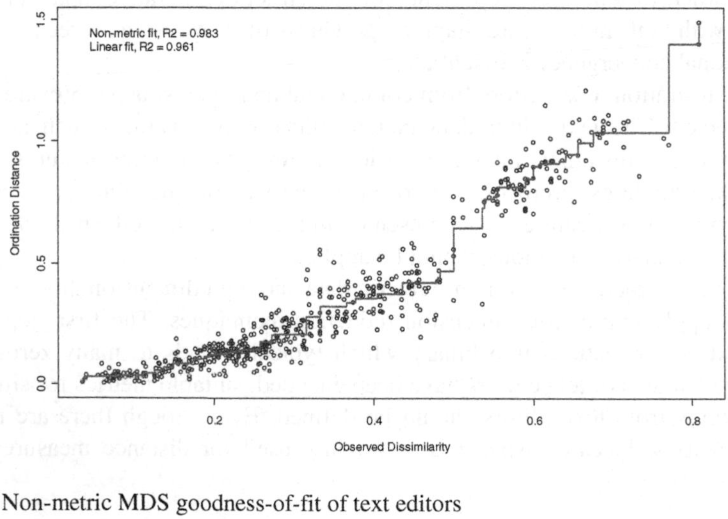

Fig. 5 Non-metric MDS goodness-of-fit of text editors

The non-metric MDS goodness-of-fit is evaluated through Shepard diagram and R-square. The Fig. 5 shows acceptable goodness-of-fit measures and visual heuristics.

The Jaccard distance measures are used as an input to cluster analysis and the Wald criteria cluster analysis indicates three features clusters as shown in Fig. 6. Dendrogram, a tree diagram used to illustrate clustering results, visualizes feature clusters in a different perspective. Particularly the heights of the dendrogram indicate relative distances from common feature centers which can provide valuable information for commonality and variability decisions.

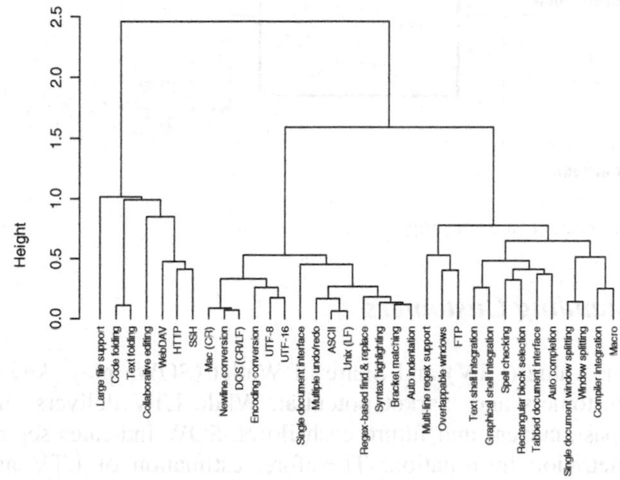

Fig. 6 Text editors Dendrogram

5 A Quantitative Product Line Scoping

Setting the appropriate scope of the product line shown in Fig. 7 is the first step to establishing initiatives in software product line. Product line scoping as multiple inputs needs strategic directions, market and customer segmentation grid, product portfolio structures. Given these inputs, suitable processing procedures such as internal and external preference mapping are required. The processing functions generate product offerings as outputs to target customer grids.

Understanding customer/market, strategic direction, and product portfolio is a cornerstone for startup scope of the product line. The business strategic direction can be reflected through product development map [43], market segmentation grid [31], and spatial-generational strategy [29]. The customer and market requirements are captured through lifetime value (LTV) and share-of-wallet (SOW). The structure of product portfolio can be revealed through multidimensional scaling (MDS) and non-metric clustering because software product features are multidimensional and non numeric.

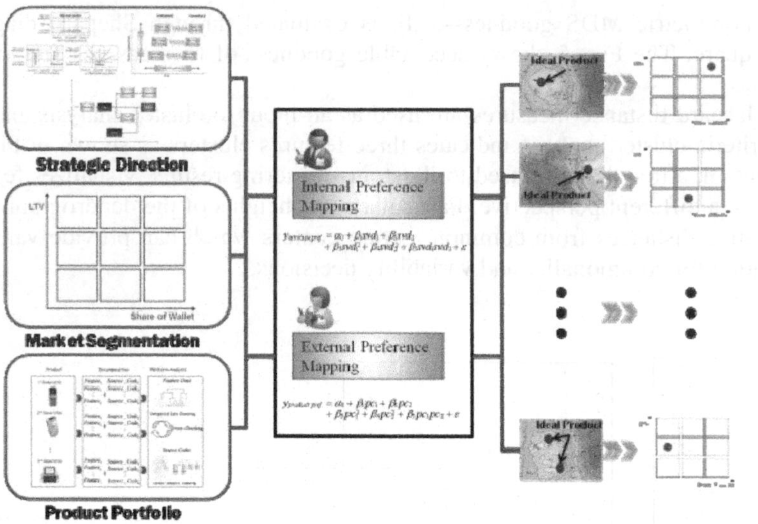

Fig. 7 Quantitative product line scoping

5.1 Understanding Customers

Customer lifetime value (LTV) and Share of Wallet (SOW) plays key roles for understanding customers and market potential. While LTV delivers customers' value based on past, current and future cash flows, SOW indicates segments potentials and penetration information. Therefore, estimation of LTV and SOW provides a cornerstone for aggregating customers into a reasonable number of customer groups and two-dimensional display of LTV and SOW shown in Fig. 8 gives insight on relative market penetration information. High LTV and low SOW

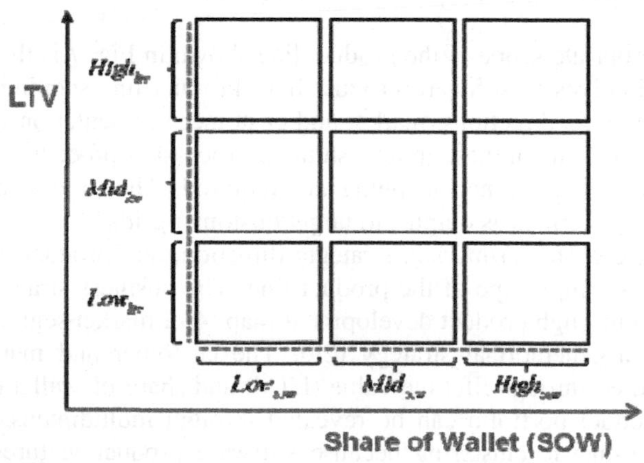

Fig. 8 Conceptual segmentation grid of LTV and SOW

reflect concerns about low market penetration and future cash cows at the same time.

Each grid, so-called market segment, reflects unique customers' profiles. Collecting customers' information and interviewing customers from the target sampling frame enables to develop customized products to the specific customer segments, which tends to increase target segments penetration and share of wallets. In addition, linking the LTV × SOW grid with leveraging strategy can reinforce the effectiveness of product line scoping.

5.1.1 Customer Lifetime Value (LTV)

Customer segmentation is defined as forming the largest group of homogeneous customers for target marketing campaigns. Customer lifetime value (LTV) is an invaluable metric for effective customer segmentation. When calculating LTV, many factors should be considered and domain experts' involvement is essential.

Regardless of LTV model configurations, an LTV model has three main components –customer's value over time, customer's length of service and a discounting factor [2,12,14,33].

$$LTV = -AC_0 + \sum_{t=0}^{n} \frac{(1-p_t)(R_t - C_t)}{(1+d)^t},$$

where AC is acquisition cost, p_t churn risk, R_t revenue during retention, C_t cost, d discount factor respectively. This basic model can be modified to meet the specific objectives. For instance, revenue cash flow can be plugged in the LTV equation and past profit contribution can be taken into account.

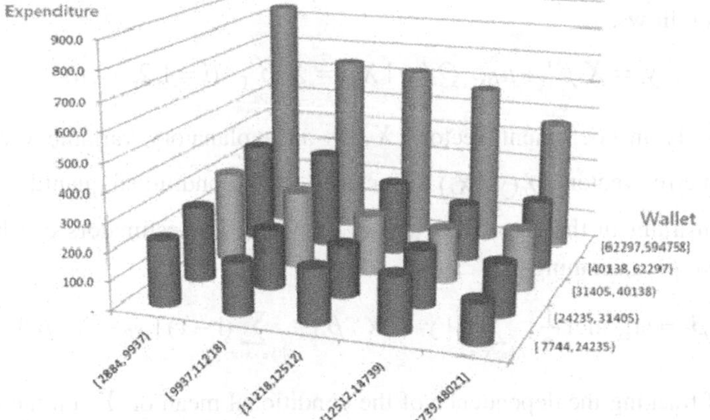

Fig. 9 Expenditure by LTV and wallet

LTV provides a good measure for past and current contributions whereas wallet estimates can be a potential market measure. The 3D bar graph on which LTV, size of wallet and expenditure are displayed at x-axis, y-axis, and z-axis as shown in Fig. 9. Analysis of market segments based on two measures gives greater insight into customer bases. To meet specific segment grid, samples satisfying target grid demographics are drawn and their needs are elicited and reflected through statistical methodologies.

5.1.2 Wallet Estimation

Share of wallet (SOW) analysis contributes to understanding the share of business a company occupies from target customers. Estimating share of wallet is quite useful for identifying opportunities in presales/marketing and product design stage.

In the credit card industry, the information of customer wallets can be obtained from credit bureau directly. Even though the share-of-wallet information cannot be obtained easily in most cases, there are two approaches for obtaining wallet estimates; top-down and bottom-up. The top-down approach is useful for business-to-business (B2B) to obtain wallets size whereas the bottom-up approach collects the wallet at the customer level, builds wallet estimation models and generalizes it to obtain actual wallet information [33]. Quantile-based estimation models can be an important tool for wallet estimation.

Quantile regression [21,22,23] is a statistical technique for estimating models for conditional quantile functions. As opposed to classical linear regression methods which minimize sums of squared residuals for conditional mean functions estimation, quantile regression methods facilitate robust estimations to the assumptions of least squares such as best linear unbiased estimation. Quantile regression generalizes the concept of an univariate quantile to a conditional quantile given one or more variables. The standard quantile regression model can be described as follows.

$$y_i = X_i\beta'_\tau + u_{\tau i}, \quad Q_\tau(y_i \mid X_i) = X_i\beta'_\tau \quad (i = 1, 2, \cdots, n)$$

where β_τ is an coefficient vector, X_i is an explanatory variable vector, and $u_{\tau i}$ is an error vector. $Q_\tau(y_i \mid X_i)$ represents τ^{th} conditional quantile and there is one constraint in the error term $Q_\tau(u_{\tau i} \mid X_i) = 0$. The estimators can be found using linear programming [3].

$$\beta_\tau = \arg\min_\beta \frac{1}{n}\{ \sum_{i: y_{\tau i} \geq X_i\beta'_\tau} \tau \mid y_{\tau i} - X'_i\beta \mid + \sum_{i: y_{\tau i} \geq X_i\beta'_\tau} (1-\tau) \mid y_{\tau i} - X'_i\beta \mid \}$$

Instead of tracking the dependence of the conditional mean of Y on the explanatory variable X, the quantile regression estimator provide a more complete description of how the conditional distribution of Y depends on X.

In the Fig. 10, increases in expenditure at the higher quantiles have been much greater than at lower quantiles and size of wallet is negatively correlated with prospect size. As opposed to ordinary least square regression, quantile regressions provide snapshots at different quantiles for measuring the impact of independent variables.

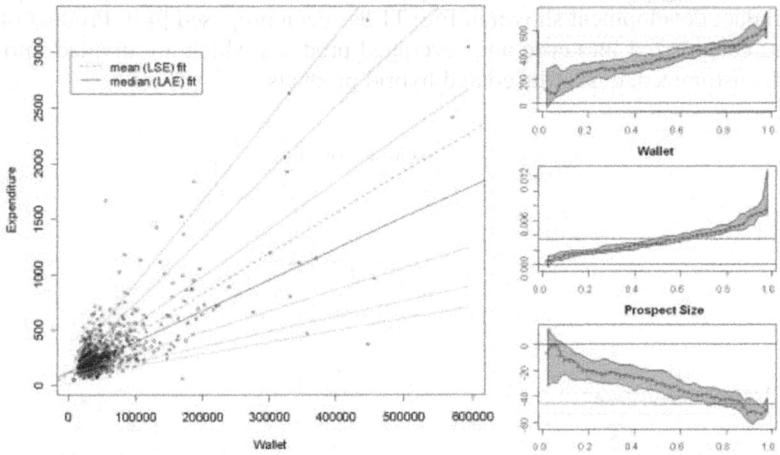

Fig. 10 Quantile regression at 5%, 10%, 25%, 75%, 90% and 95% quantiles

5.2 *Strategic Direction*

Depending on SPL leveraging strategy, the opportunistic reuse can make differences. Appropriately aligning segmentation strategy with platform leveraging strategy can bring opportunities to magnify reuse benefits. The customer and

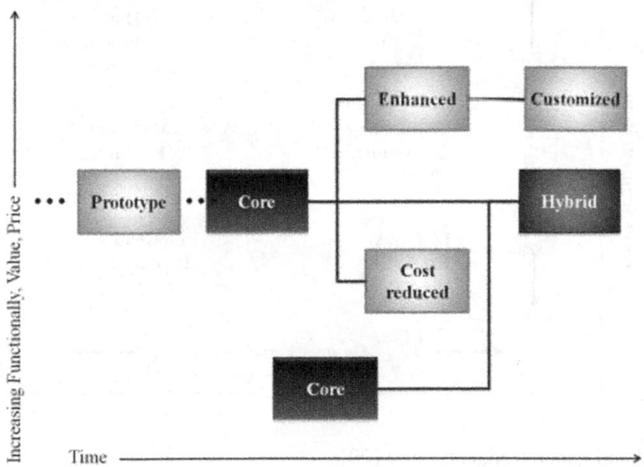

Fig. 11 Product development map [43]

product requirements for SPL scoping are captured through LTV/SOW and MDS, respectively. The business strategy can be reflected through product development map, market segmentation grid, and spatial and generational strategy.

The strategy for constantly introducing new products into markets has been developed in accordance with the managerial directions. Product map concept for new product development shown in Fig. 11 has been proposed [43]. Product offerings are categorized into core and leveraged products which are divided into enhanced, customized, cost reduced and hybrid products.

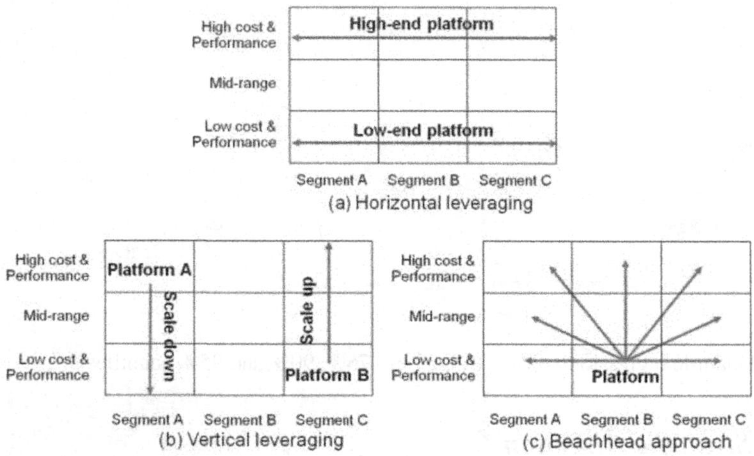

Fig. 12 Market segmentation grid and platform leveraging strategies [31]

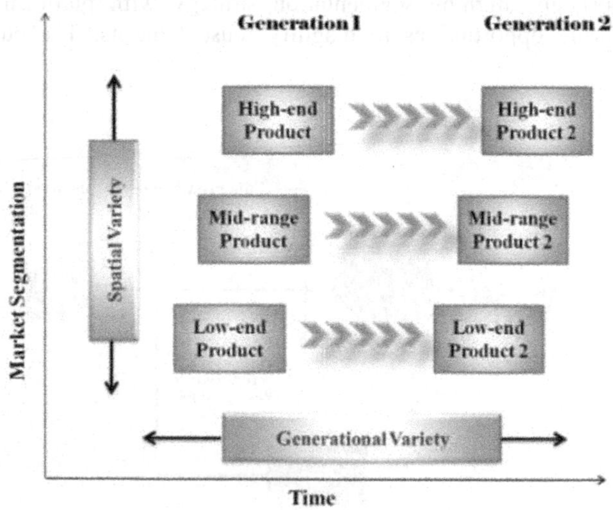

Fig. 13 Spatial and generational strategy [29]

Market Segmentation Grid and platform leveraging strategy [31] are articulated in Fig. 12. The horizontal axis is market segments and the vertical axis is cost & performance. If the same platform is used to cover multiple segments, then this is the case of the horizontal platform leveraging strategy. If the single platform is implemented from a low to a high end of the market segment, then this is the case of the vertical leveraging strategy. The beachhead strategy combines the vertical leveraging strategy with horizontal leveraging strategy.

The platform leveraging strategy based on spatial and generational variety shown in Fig. 13 has been proposed [29]. Combining product development map with market segmentation grid provides insight into designing effective and efficient platform.

5.3 Reflecting Voices of Engineers and Customers

Internal preference analysis and external preference analysis are used to reflect voices of engineers and customers through multivariate regression models as shown in Fig. 14. Statistical techniques such as singular value decomposition (SVD) and principle component analysis (PCA) are applied for low dimensional independent variables on which engineers/customers' preference as dependent variable is regressed.

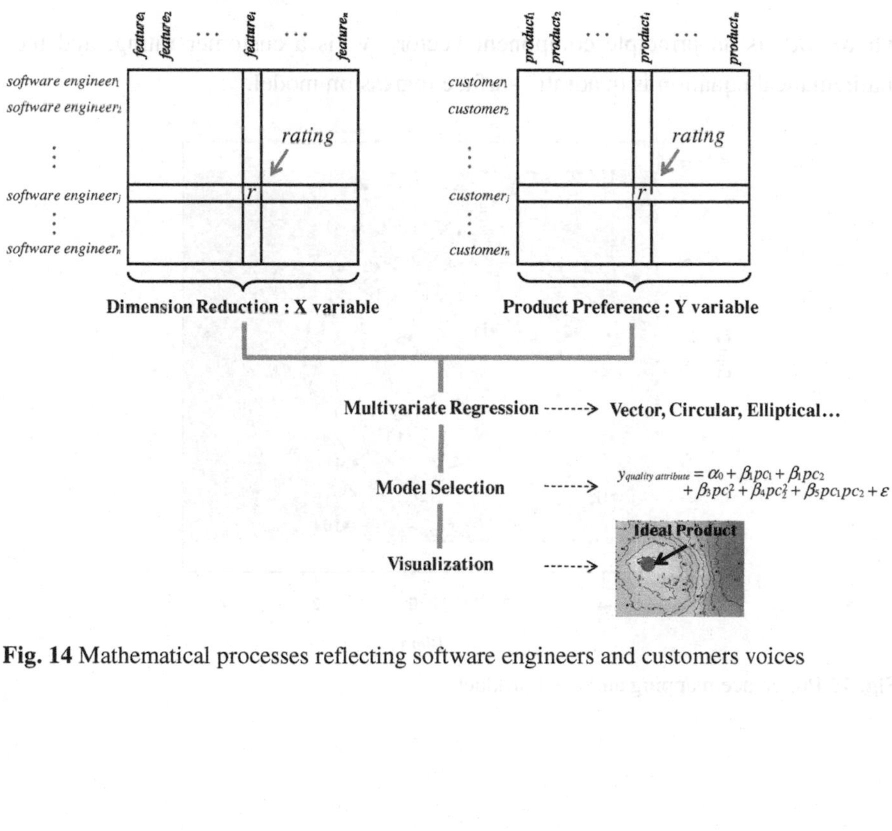

Fig. 14 Mathematical processes reflecting software engineers and customers voices

Understanding what customers think and want needs preference and perception information. Although consumers can state products preferences explicitly, they cannot provide preference mechanism for products choice. Preference mapping techniques relate products and features to customer preference or perception.

Customers reveal their preferences through rating products or attributes. Target customers are drawn from market or customer segmentation grids and preferences on products and features are collected. Internal preference analysis emphasizes consumer preferences and hence captures more of consumer understanding while external preference analysis works from perceptual or sensory information and hence captures more of product understanding [20].

Statistical techniques such as singular value decomposition (SVD) and principle component analysis (PCA) are applied for dimension reduction, low-dimension interpretation, and independent variables for multivariate regression. Regression analysis with variable selection is performed to incorporate external information and build robust regression models for identifying customers' perception and developing ideal products. The principal components for dimension reduction are extracted and variable selection methods such as forward, backward, stepwise, best-subset, and Akaike information criteria (AIC) and Bayesian Information criteria (BIC) are applied for obeying principle of parsimony. Fitting robust regression models to modeling data leads to model assessment and explanation.

$$y_{\text{product preference}} = \alpha_0 + \beta_1 pc_1 + \beta_2 pc_2 + \beta_3 pc_1^2 + \beta_4 pc_2^2 + \beta_5 pc_1 pc_2 + \varepsilon$$

where pc_i is an principle component vector, y_i is a customer rating, and the mathematical equation is quadratic surface regression model.

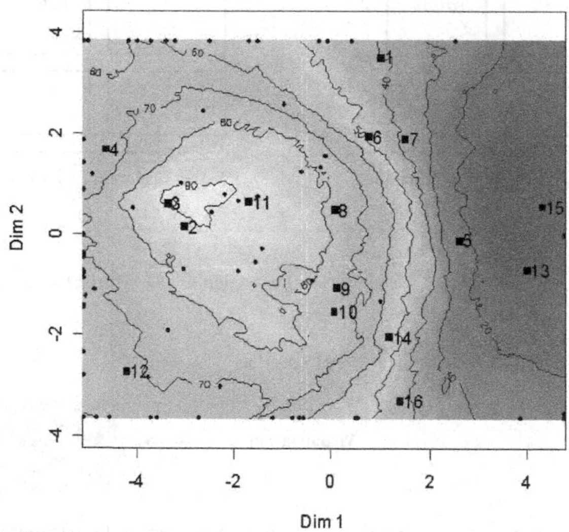

Fig. 15 Preference mapping and ideal product

Dimension reduction methods and linear regression with variable selections are applied to reflect voices of engineers and customers. Regressing preference data onto a product space contributes to revealing ideal products and features. Preference mapping provides a powerful approach to designing attractive products listening to voices of customers and engineers as shown in Fig. 15. Products 2, 3, and 11 are close to an ideal product and products with low utility features are displayed at the same time. Examination of products close to an ideal point with marketing grids strategies helps software engineers to manufacture what the market wants.

5.4 Offering Mass Customized Products to Target Segments

Target customers' preferences are elicited through internal and external preference analysis such that customized products with ideal features are ready to design to penetrate the niche market grid. When building multivariate regression models, application of variable selection techniques for vector model, circular ideal point model, elliptical ideal point model, and quadratic surface model contributes to revealing preferred products with ideal feature sets for specific segments. By regressing customer's preference ratings on coordinates of low dimensions from singular value decomposition and principle component analysis, an explanation can be made of relating customers to products.

The customized products embedding segments preferences are offered to target segments which represent homogeneous groups and niche potentials as shown in Fig. 16.

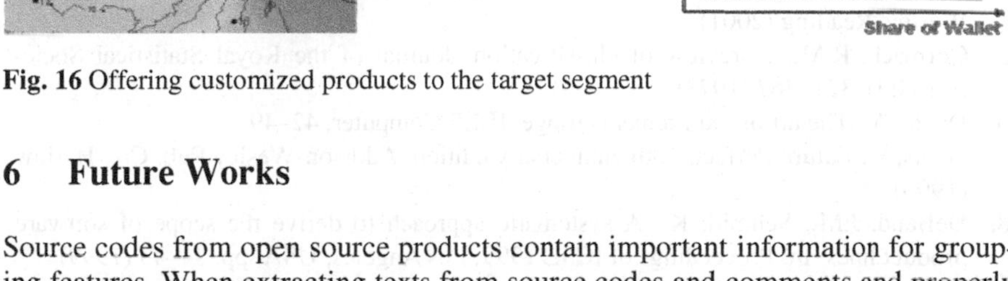

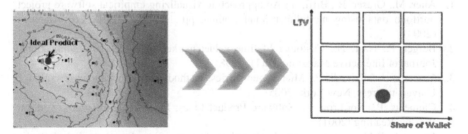

Fig. 16 Offering customized products to the target segment

6 Future Works

Source codes from open source products contain important information for grouping features. When extracting texts from source codes and comments and properly processing textual data through ETL (Extraction, Transformation and Loading) processes, information extraction methods can reveal hidden structures for asset scoping. Latent semantic indexing which uses singular value decomposition for dimension reduction can show bottom-up asset scoping views.

By comparing scoping results from feature-level grouping with source-code level scoping outputs from latent semantic indexing analysis, the accuracy of asset scoping can be improved and the MDS and non-metric clustering asset scoping can be validated from the bottom-up approach as well.

7 Conclusion

Many recent researches have discussed the importance of determining appropriate scope of product line and linking market information, and developing customized products for market segments, but few have offered a logically integrated and tailored framework embracing modern market analysis, mass customization strategies, and advanced statistical methods.

We proposed a market-driven quantitative scoping framework. Modern market analysis such as LTV and SOW are introduced to reflect customers' needs correctly and estimate market size and values. Multidimensional scaling (MDS) and non-metric clustering became useful tools to visualize commonality and variability features by facilitating communications among stakeholders to reveal product portfolio structures quickly. The text editor case study demonstrated applicability of the method for determination of commonality and variability. Mass customization leveraging strategies and preference mapping contributed to reflecting the management and engineers' voices and offering customized products to market segments.

References

1. Auer, M., Graser, B., Biffl, S.: An approach to visualizing empirical software project portfolio data using multidimensional scaling, pp. 504–512. IEEE, Los Alamitos (2003)
2. Berger, P., Nasr, N.: Customer lifetime value: marketing models and applications. Journal of Interactive Marketing 12(1) (1998)
3. Cameron, C., Trivedi, P.: Microeconometrics: methods and applications. Cambridge University Press, New York (2005)
4. Clements, P., Northrop, L.: Software Product Lines: Practices and Pattern. Addison Wesley, Reading (2001)
5. Cormack, R.M.: A review of classification. Journal of the Royal Statistical Society 134(3), 321–367 (1973)
6. Davis, A.: The art of requirements triage. IEEE Computer, 42–49
7. Davis, S.: Future Perfect, 10th anniversary edition. Addison-Wesley Pub. Co., Harlow (1996)
8. DeBaud, J.M., Schmid, K.: A systematic approach to derive the scope of software product lines. In: Proceedings of ICSE 1999, Los Angeles, USA, pp. 34–43 (1999)
9. Fraley, C., Raftery, A.: Model-based clustering, discriminant analysis and density estimation. Journal of the American Statistical Association 97(458), 611–631 (2002)
10. Helferich, A., Herzwurm, G., Schockert, S.: QFD-PPP: Product line portfolio planning using quality function deployment. In: Obbink, H., Pohl, K. (eds.) SPLC 2005. LNCS, vol. 3714, pp. 162–173. Springer, Heidelberg (2005)

11. Helferich, A., Schmid, K., Herzwurm, G.: Reconciling marketed and engineered software product lines. In: 10th International Software Product Line Conference, SPLC 2006 (2006)
12. Helsen, K., Schmittlein, D.: Analyzing duration times in marketing: evidence for the effectiveness of hazard rate models. Marketing Science 11(4), 395–414 (1993)
13. Jaccard, P.: Distribution de la flore alpine dans le basin des Dranes et dans quelques regions voisines. Bulletin del la Société Vaudoise des Sciences Naturelles 37, 241–272 (1901)
14. Jain, D., Singh, S.: Customer lifetime value research in marketing: a review and future directions. Journal of Interactive Marketing 16(2), 34–46 (2002)
15. Jiao, J.R., Simpson, T.W., et al.: Product family design and platform-based product development: a state-of-the-art review. Journal of Intelligent Manufacturing 18, 5–29 (2006)
16. John, I., Knodel, J., Lehner, T., Muthig, D.: A practical guide to product line scoping. In: 10th International Software Product Line Conference, SPLC 2006 (2006)
17. Kang, K.C., Donohoe, P., Koh, E., Lee, J.J., Lee, K.: Using a marketing and product plan as a key driver for product line asset development. In: Chastek, G.J. (ed.) SPLC 2002. LNCS, vol. 2379, pp. 366–382. Springer, Heidelberg (2002)
18. Kaufman, L., Rousseeuw, P.J.: Finding Groups in Data: An introduction to Cluster Analysis. John Wiley & Sons, New York (1990)
19. Kaul, A., Rao, V.R.: Research for product positioning and design decisions: an integrative review. International Journal of Research in Marketing 12, 293–320 (1995)
20. Kleef, E., Trijp, H., Luning, P.: Internal versus external preference analysis: an exploratory study on end-user evaluation. Food Quality and Preference 17, 387–399 (2006)
21. Koenker, R.: Quantile regression, pp. 222–228. Cambridge Press, Cambridge (2005)
22. Koenker, R., Bassett, G.: Regression quantiles. Econometrica 46(1), 33–50 (1978)
23. Koenker, R., Hallock, K.: Quantile regression. Journal of Economic Perspectives 15(4), 143–156 (2001)
24. Krishnan, V., Ulrich, K.T.: Product development decisions: a review of the literature. Management Science 47, 1–21 (2001)
25. Kruskal, J.B.: Multidimensional scaling by optimizing goodness of fit to a nonmetric hypothesis. Psychometrika 29, 1–27 (1964)
26. Kruskal, J.B.: Multidimensional scaling: a numerical method. Psychometrika 29, 1–27 (1964)
27. Lee, K., Kang, K.C., Lee, J.: Concepts and Guidelines of Feature Modeling for Product Line Software Engineering. In: Proceedings of the 7th Reuse Conference, Austin, April 15-19, pp. 62–77 (2002)
28. MacQueen, J.: Some methods for classification and analysis of multivariate observations. In: Proceedings of the 5th Berkeley Symposium on Math. Stat. and Prob., Statistics, vol. I (1967)
29. Martin, M., Ishii, K.: Design for Variety: A Methodology for Developing Product Platform Architectures. In: Proceedings of DETC 2000, Baltimore Maryland, pp. 10–13 (2000)
30. Mead, A.: Review of the development of multidimensional scaling methods. The Statistician 41(1), 27–39 (1992)
31. Meyer, M.H., Lehnerd, A.P.: The power of product platforms. The Free Press, New York (1997)
32. Meyer, M.H., Tertzakian, P., et al.: Metrics for managing research and development in the context of the product family. Management Science 43 (1), 88–111 (1997)

33. Rosset, S., Neumann, E., Eick, U., Vatnik, N.: Customer lifetime value for decision support. In: Data Mining and Knowledge Discovery, pp. 321–339 (2003)
34. Schmid, K.: A comprehensive product line scoping approach and its validation. In: Proceedings of ICSE 2002, Orlando, USA, pp. 593–603 (May 2002)
35. Shepard, R.N.: The analysis of proximities: multidimensional scaling with an unknown distance function, I. Psychometrika 27, 125–140 (1962)
36. Shepard, R.N.: The analysis of proximities: MDS with an unknown distance function, II. Psychometrika 27, 219–246 (1962)
37. Simpson, T.W.: Product Platform Design and Customization: Status and Promise. Artificial Intelligence for Engineering Design, Analysis, and Manufacturing 18(1), 3–20 (2004)
38. Standish Group: Chaos: A Recipe for Success. Standish Group International (1999)
39. Takane, Y., Young, F.W., de Leeuw, J.: Non-metric individual differences MDS: an alternating least squares method with optimal scaling features. Psychometrika 42, 7–67 (1978)
40. Torgerson, W.S.: Multidimensional scaling: I. Theory and method. Psychometrika 17, 401–419 (1952)
41. Van Zyl, J., Walker, A.: Product line balancing-an empirical approach to balance features across product lines. In: IEEE Engineering Mgmt. Conference Proceedings, pp. 1–6. IEEE Electronic Library (2001)
42. Verner, J.M., Evanco, W.M.: In-house software development: what project mgmt practices lead to success? IEEE Software (2005)
43. Wheelwright, S., Sasser, W.E.: The New Product Development Map. Harvard Business Review, 112–125 (1989)
44. Williams, W.T.: Principles of clustering. Annual Review of Ecology and Systematics 2, 303–326 (1971)

Clustering and Analyzing Embedded Software Development Projects Data Using Self-Organizing Maps

Kazunori Iwata, Toyoshiro Nakashima, Yoshiyuki Anan, and Naohiro Ishii

Abstract. In this paper, we cluster and analyze data from the past embedded software development projects using self-organizing maps (SOMs)[9] that are a type of artificial neural networks that rely on unsupervised learning. The purpose of the clustering and analysis is to improve the accuracy of predicting the number of errors. A SOM produces a low-dimensional, discretized representation of the input space of training samples; these representations are called maps. SOMs are useful for visualizing low-dimensional views of high-dimensional data, a multidimensional scaling technique. The advantages of SOMs for statistical applications are as follows: (1) data visualization, (2) information processing on association and recollection, (3) summarizing large-scale data, and (4) creating nonlinear models. To verify our approach, we perform an evaluation experiment that compares SOM classification to product type classification using Welch's t-test for Akaike's Information Criterion (AIC). The results indicate that the SOM classification method is more contributive

Kazunori Iwata
Department of Business Administration, Aichi University, 370 Shimizu, Kurozasa-cho, Miyoshi, Aichi, 470-0296, Japan
e-mail: `kazunori@vega.aichi-u.ac.jp`

Toyoshiro Nakashima
Department of Culture-Information Studies, Sugiyama Jogakuen University, 17-3 Moto-machi, Hoshigaoka, Chikusa-ku, Nagoya, Aichi, 464-8662, Japan
e-mail: `nakasima@sugiyama-u.ac.jp`

Yoshiyuki Anan
Base Division, Omron Software Co., Ltd., Higashiiru, Shiokoji-Horikawa, Shimogyo-ku, Kyoto, 600-8234, Japan
e-mail: `y-anan@mx.omronsoft.co.jp`

Naohiro Ishii
Department of Information Science, Aichi Institute of Technology, 1247 Yachigusa, Yakusa-cho, Toyota, Aichi, 470-0392, Japan
e-mail: `ishii@aitech.ac.jp`

R. Lee (Ed.): Software Eng. Research, Management & Appl. 2011, SCI 377, pp. 47–59.
springerlink.com

than product type classification in creating estimation models, because the mean AIC of SOM classification is statistically significantly lower.

Keywords: Self-organizing maps, clustering, embedded software development.

1 Introduction

In this paper, we cluster and analyze data from the past embedded software development projects using self-organizing maps (SOMs) [9].

Growth in the information industry has caused a wide range of uses for information devices and the associated need for more complex embedded software that provides these devices with the latest performance and function enhancements [5, 12]. Consequently, it is increasingly important for embedded software development corporations to develop software efficiently, guarantee delivery times and quality, and keep development costs low [3, 10, 11, 13, 14, 16, 17, 18]. Hence, companies and divisions involved in the development of such software are focusing on various types of improvements, particularly on process improvement. Predicting manpower requirements for new projects and guaranteeing software quality are especially important, because these predictions directly relate to cost, while the quality reflects on the reliability of the corporation.

In the field of embedded software, development techniques, management techniques, tools, testing techniques, reuse techniques, and real-time operating systems have already been studied [4, 11, 13]; however, there is little research conducted on the scale of development efforts, in particular the relationship between the number of efforts and the number of errors, based on the data accumulated from past projects. In our previous work, we described the prediction of total efforts and errors using an artificial neural network (ANN) [6, 7] . In addition, we improved the ANN models using a new normalization method [8]; however, the data include exceptions that cause the model to be less effective and potentially inaccurate. Therefore, we propose clustering the data to discover and avoid such exceptions.

In this paper, we present our work of analyzing and clustering the data using a SOM. We perform an evaluation experiment that compares this SOM classification to product type classification [7] using Welch's t-test[19]. The comparison results indicate that SOM classification is more contributive in creating estimation models than product type classification.

In addition to this introductory section, our paper is organized as follows: section 2 provides the data sets to be used for clustering; section 3 summarizes SOMs; section 4 presents our experimental methods and results; and section 5 provides our conclusions and future work.

2 Data Sets for Clustering

We aim to cluster the following data in order to analyze the relevance among software development projects:

Err: The number of errors, which is specified as the total number of errors for an entire project.

$Type_n$: The machine type equipped with the developed software. Here, n indicates the type and ranges from 1 to 13. One of them is 1 and the others are 0 (shown in Table 1).

$Language_n$: The programing language used to build the software of the project. Here, n indicates the language and ranges from 1 to 7. If a project uses the $Language_1$ and $Language_3$, then $Language_1$ and $Language_3$ indicate 1 and the others indicate 0 (shown in Table 2).

$Leader_n$: The project leader. Here, n indicates the type and from 1 to 24. One of them is 1 and the others are 0 (shown in Table 2).

V_{new}: Volume of newly added steps, which denotes the number of steps in the newly generated functions of the target project.

V_{modify}: Volume of modification, which denotes the number of steps modified and added to the existing functions to use the target project.

V_{survey}: Volume of original project, which denotes the original number of steps in the modified functions; the number of steps deleted from the functions can therefore be calculated.

V_{reuse} : Volume of reuse, which denotes the number of steps in functions of which only an external method has been confirmed and which are applied to the target project design without confirming the internal contents.

Tables 1, 2 and 3 show data example.

3 Basis for Using Self-Organizing Maps

3.1 Basic Self-Organizing Maps

A SOM can produce a low-dimensional, discretized representation of the input space of training samples. These representations are called maps. SOMs are useful for visualizing low-dimensional views of high-dimensional data, a multidimensional scaling technique. They are a type of ANNs that are trained using unsupervised learning; however, they differ from other ANNs in the sense that they use neighborhood functions to preserve the topological properties of the input space. Like most ANNs, they operate in the following two modes: training and mapping.

Training builds the map using input examples with known results (i.e., training data). It is a competitive process, also called vector quantization.

Mapping automatically classifies a new input vector and produces a result.

SOMs consist of components called nodes or neurons. Associated with each node is a weight vector with the dimension same as the input data vectors and a position in the map space. The usual arrangement of nodes is a regular spacing in a hexagonal or rectangular grid. A SOM describes a mapping from a higher-dimensional input space to a lower-dimensional map space. The procedure for placing a vector from the input space to the map space is to find a node with the weight vector closest to

Table 1 Data Example 1

Project No.	$Type_1$	$Type_2$	$\cdots$	$Type_{12}$	$Type_{13}$	$Language_1$	$Language_2$	$\cdots$	$Language_6$	$language_7$
1	1	0	$\cdots$	0	0	1	1	$\cdots$	0	0
2	1	0	$\cdots$	0	0	1	1	$\cdots$	0	0
					⋮					
7	0	1	$\cdots$	0	0	0	1	$\cdots$	0	0
					⋮					

Table 2 Data Example 2

Project No.	Err	$Leader_1$	$Leader_2$	$\cdots$	$Leader_{23}$	$Leader_{24}$
1	0.457945512	1	0	$\cdots$	0	0
2	0.244804866	0	1	$\cdots$	0	0
			⋮			
7	0.648149451	0	0	$Leader_3 = 1$	0	0
			⋮			

Table 3 Data Example 3

Project No.	V_{new}	V_{modify}	V_{survey}	V_{reuse}
1	0.187309827	0.29864242	0.26730765	0.51464728
2	0.173525764	0.17746571	0.20254854	0.25550976
		⋮		
7	0.611294395	0.64299185	0.34650809	0.16160979
		⋮		

the vector in the data space and then assign the map coordinates of this found node to the vector.

The learning algorithm is as follows:

1. The weights of the neurons are initialized to small random values.
2. A input vector $x(t)$ is given to the network.
3. The neuron with the weight vector closest to that of the input $x(t)$ is identified as the best matching unit (BMU); to find the BMU, Euclidean distances to all neuron are used.
4. The weights of the BMU and its neighborhood units are updated. The magnitude of the change decreases with time and with distance from the BMU.
   The update formula for a neuron with weight vector $m_i(t)$ is

$$m_i(t+1) = m_i(t) + h_{ci}(t)\alpha(t)[x(t) - m_i(t)], \tag{1}$$

   where $\alpha(t)$ is a monotonically decreasing learning coefficient and $x(t)$ is the input vector and $h_{ci}(t)$ is the neighborhood function. The neighborhood function

$h_{ci}(t)$ depends on the lattice distance between the BMU and neuron i. In its simplest form, $h_{ci}(t)$ is one for all neurons sufficiently close to the BMU and zero for all others.

5. This process is repeated (by going to step 2) for each input vector for a (usually large) number of cycles λ.

3.2 Spherical Self-Organizing Maps

Unlike Basic SOMs (described above) in which the output space is two-dimensional, spherical SOMs consist of a spherical output space.

Fig.1 shows the results of Basic SOM applied to sample data. In Fig.1, black and white dots are nodes. If the Euclidean distance between nodes is smaller, the color is closer to white; otherwise, the color is closer to black.

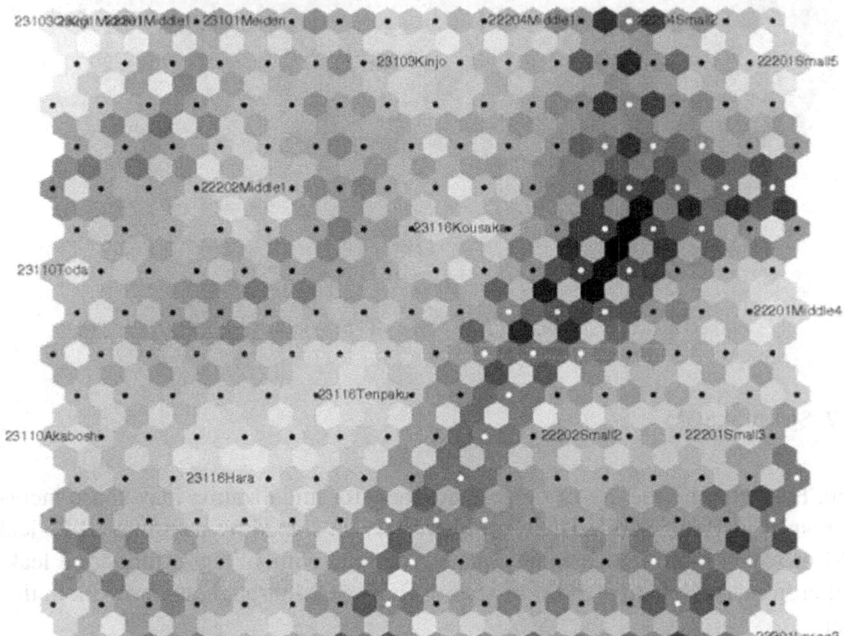

Fig. 1 Basic SOM

Fig.2 shows an example of a Spherical SOM in which white colored hexagons are nodes. The main difference between Basic SOM and Spherical SOM is the absence of edges in Spherical SOM.

Because the size of the learning domain is limited in Basic SOM, there is the possibility of information leak during the learning process in the two-dimensional

Fig. 2 Spherical SOM

plane. Furthermore, the use of random numbers for initialization may cause incorrect results in Basic SOM. These problems can be solved by using the Spherical SOM in which multi-dimensional data can be read without any information leak. We therefore use Spherical SOM instead of Basic SOM throughout our work in this paper.

3.3 Advantage of Self-Organizing Maps

The advantages of SOMs for statistical applications include the following:

- Data visualization.

  A SOM is used to visualize the original data or the results of statistical analysis. In addition, it is useful for finding the nonlinear complex relationship among attributes of data.

- Information processing on association and recollection
  SOMs enable reasonable inferences to be made from incomplete information via association and recollection.
- Summarizing large-scale data
  By using a discretized representation of the input space of training samples, a SOM acts as a form of data compression.
- Creating nonlinear models
  Each node of a SOM represents a linear model. The nodes of the SOM place important data spaces, and the SOM can represent nonlinear features of the data spaces. Therefore, SOMs can translate original statistical analysis models into nonlinear models.

4 Evaluation Experiments

4.1 Results of Clustering

We clustered 149 embedded software development projects, which contain the data shown in Section 2, using Spherical SOM, the results of which are shown in Fig.3. The data are numbered from 1 to 149. Although the figure illustrates the relationship between the projects, it is difficult to understand the entire picture. The results are therefore further shown as dendrograms in Figs.4 and 5. In these figures, projects with a similar nature are located close together. To better see this phenomenon, Fig.6 shows a subset of Fig.4 in more detail, showing projects numbered 17 and 23 as being the closest projects. The same relationships are found among clusters with project numbers 115, 122, and 123, as well as 118, 119, and 124. We define cluster size by first selecting a node point; for example, if the circled node in Fig.6 is selected, project numbers 1, 21, 5, 17, and 23 all belong to the same cluster.

4.2 Comparison Experiment

We performed a comparison experiment to confirm our clustering results obtained via Spherical SOM. Our method of confirmation is as follows:

1. Set up two types of classification data: (1) one set based on product type classification (mimicking [7], shown below in Table 4) and (2) a second set in conformity to the dendrograms shown in Figs.4 and 5. The number of classes is fixed at five. In general, SOM classification can have more than five classes, but we fix this at five for a better comparison. Results of classification by SOM are summarized in Table 5.
2. Create error estimation models for each class by using multiple regression analysis in which a best subset selection procedure is used to select explanatory variables based on Akaike's Information Criterion (AIC) [1] . The smaller the AIC value, the higher the accuracy or fit of the model.
3. Compare the mean AIC of each classified data set.

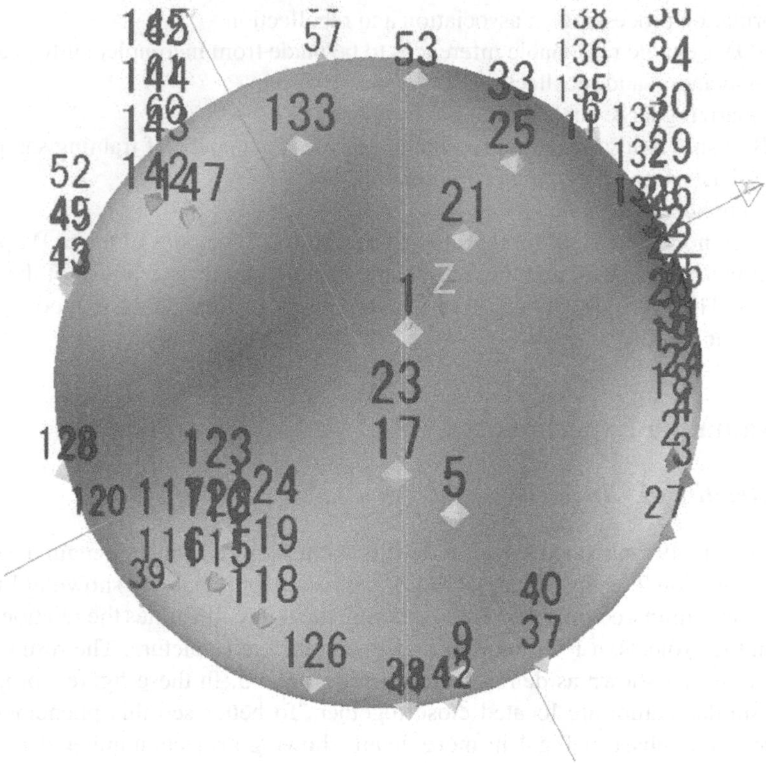

Fig. 3 Results of SOM Clustering

Results of this experiment are summarized in Table 6.

Next, we compare the accuracy of our results using Welch's t-test [19]. The t-test (called Student's t-test) [15] is used as a test of the null hypothesis that the means of two normally distributed populations are equal. Welch's t-test is used when the variances of two samples are assumed to be different to test the null hypothesis that the means of two non-normally distributed populations are equal if the two sample sizes are equal [2]. Results of the t-test for our comparison experiment are summarized in Table 7.

The null hypothesis, in this case, is that "there is no difference between the means of AIC of product type classification and SOM classification." The results in Table 7 indicate a statistically significant difference, because the p-value is 0.07667 and thus greater than 0.05 and less than 0.1.

SOM classification surpasses the product type classification in accuracy for creating an error estimation model, because of the fewer AIC. Hence, SOM classification makes a better contribution to accurately create estimation models.

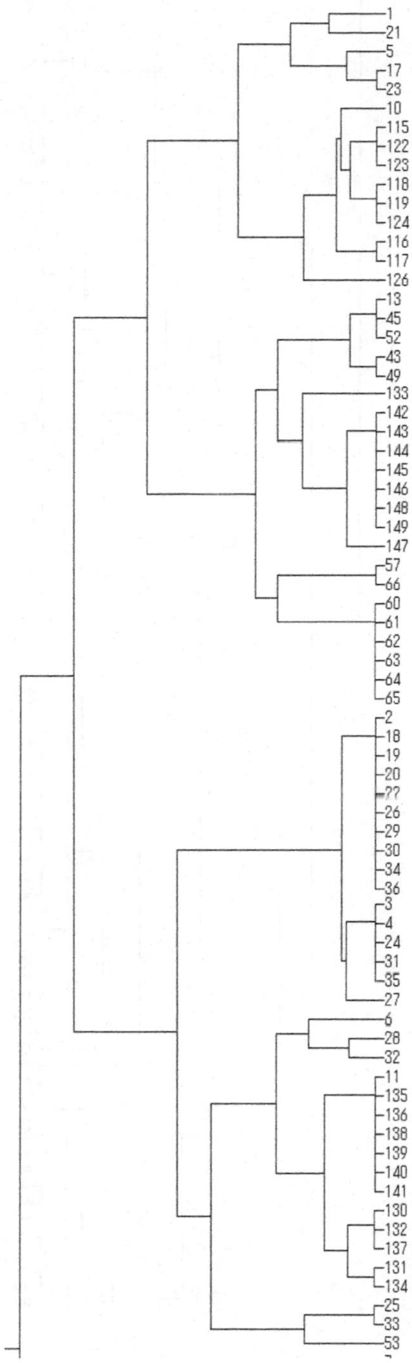

Fig. 4 Clustering Results 1

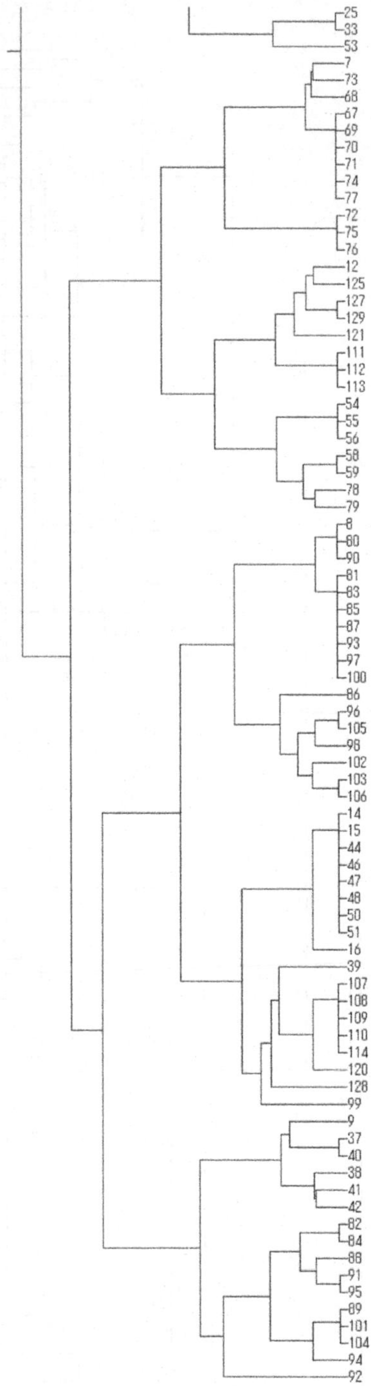

Fig. 5 Clustering Results 2

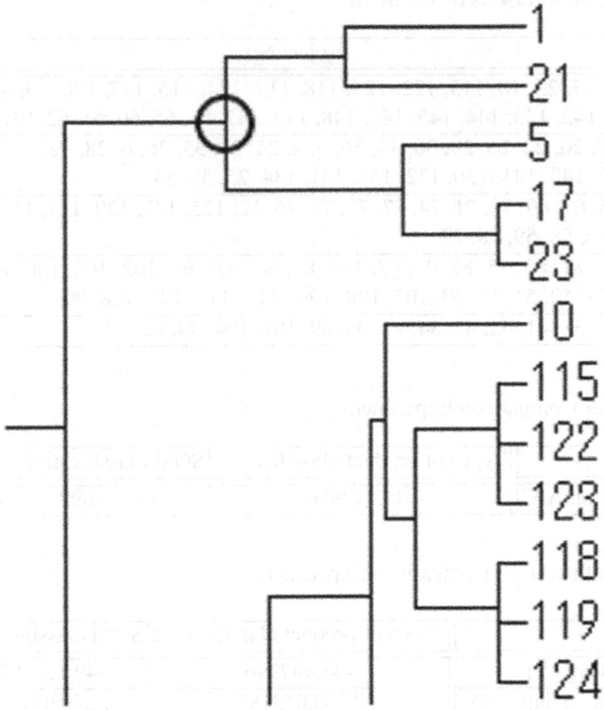

Fig. 6 Example of Clustering

Table 4 Classification Results by the Types of Product

Class	Project No.
1	1, 2, 3, 4, 5, 17, 18, 19, 20, 21, 22, 23, 24, 25, 26, 27, 28, 29, 30, 31, 32, 33, 34, 35, 36, 37, 38, 39, 40, 41, 42
2	7, 54, 55, 56, 57, 58, 59, 60, 61, 62, 63, 64, 65, 66, 67, 68, 69, 70, 71, 72, 73, 74, 75, 76. 77, 78, 79
3	8, 80, 81, 82, 83, 84, 85, 86, 87, 88, 89, 90, 91, 92, 93, 94, 95, 96, 97, 98, 99, 100, 101, 102, 103, 104, 105, 106, 107, 108, 109, 110, 111, 112, 113, 114
4	9, 10, 115, 116, 117, 118, 119, 120, 121, 122, 123, 124, 125, 126, 127, 128, 129, 11, 130, 131, 132, 133, 134, 135, 136, 137, 138, 139, 140, 141
5	12, 14, 15, 16, 43, 44, 45, 46, 47, 48, 49, 50, 51, 52, 142, 143, 144, 145, 146, 147, 148, 149

Table 5 Classification Results by the SOM

Class	Project No.
1	1, 21, 5, 17, 23, 10, 115, 122, 123, 118, 119, 124, 116, 117, 126, 13, 45, 52, 43, 49, 133, 142, 143, 144, 145, 146, 148, 149, 147, 57, 66, 60, 61, 62, 63, 64, 65
2	2, 18, 19, 20, 22, 26, 29, 30, 34, 36, 3, 4, 24, 31, 35, 26, 6, 28, 32, 11, 135, 136, 138, 139, 140, 141, 130, 132, 137, 131, 134, 25, 33, 53
3	7, 73, 68, 67, 69, 70, 71, 74, 77, 72, 75, 76, 12, 125, 127, 129, 121, 11, 112, 113, 54, 55, 56, 58, 59, 78, 79
4	8, 80, 90, 81, 83, 85, 87, 93, 97, 100, 86, 96, 105, 98, 102, 103, 106, 14, 15, 44, 46, 47, 48, 50, 51, 16, 39, 107, 108, 109,, 110, 114, 120, 128, 99
5	9, 37, 40, 38, 41, 42, 82, 84, 91, 95, 89, 101, 104, 94, 92

Table 6 Results of Comparison Experiment

–	Types of product classification	SOM classification
Mean AIC	-43.447996	-49.123698

Table 7 Results of t-test for Comparison Experiment

–	Types of product classification	SOM classification
Mean ($\bar{X}$)	-43.447996	-49.123698
Standard deviation(s)	11.76053187	25.70308074
Sample size (n)	5	5
Degrees of freedom (v)	7.027	
T-value (t_0)	2.0735	
P-value	0.07667	

5 Conclusions

In this paper, we described our approach of clustering and analyzing past embedded software development projects using Spherical SOM. We also described our evaluation experiments that compared SOM classification with product type classification using Welch's t-test. Results indicated that SOM classification is more contributive than product type classification for creating estimation models, because the mean AIC of the SOM classification is statistically significantly lower.

In our analysis, we used multiple regression analysis to create our model; more complex models should be considered in future work to improve the accuracy. Furthermore, the clustering and model creation processes are independent of one another; in the future, we aim to combine these modules for the ease of use.

In our analysis, we also created a model to predict the final number of errors; in future work, we plan to predict errors at various stages (e.g., halfway) of the development process of a project. Overall, more data are needed to further support our work.

References

1. Akaike, H.: Information theory and an extention of the maximum likelihood principle. In: Petrov, B.N., Csaki, F. (eds.) 2nd International Symposium on Information Theory, pp. 267–281 (1973)
2. Aoki, S.: In testing whether the means of two populations are different (in Japanese), http://aoki2.si.gunma-u.ac.jp/lecture/BF/index.html
3. Boehm, B.: Software engineering. IEEE Trans. Software Eng. C-25(12), 1226–1241 (1976)
4. Futagami, T.: Embedded software development tools update. Journal of Information Processing Society of Japan(IPSJ) 45(7), 704–712 (2004)
5. Hirayama, M.: Current state of embedded software. Journal of Information Processing Society of Japan(IPSJ) 45(7), 677–681 (2004) (in Japanese)
6. Iwata, K., Anan, Y., Nakashima, T., Ishii, N.: Using an artificial neural network for predicting embedded software development effort. In: Proceedings of 10th ACIS International Conference on Software Engineering, Artificial Intelligence, Networking, and Parallel/Distributed Computing – SNPD 2009, pp. 275–280 (2009)
7. Iwata, K., Nakashima, T., Anan, Y., Ishii, N.: Error estimation models integrating previous models and using artificial neural networks for embedded software development projects. In: Proceedings of 20th IEEE International Conference on Tools with Artificial Intelligence, pp. 371–378 (2008)
8. Iwata, K., Nakashima, T., Anan, Y., Ishii, N.: Improving accuracy of an artificial neural network model to predict effort and errors in embedded software development projects. In: Lee, R., Ma, J., Bacon, L., Du, W., Petridis, M. (eds.) SNPD 2010. SCI, vol. 295, pp. 11–21. Springer, Heidelberg (2010)
9. Kohonen, T.: Self-Organizing Maps, 3rd edn. Springer, Heidelberg (2000)
10. Komiyama, T.: Development of foundation for effective and efficient software process improvement. Journal of Information Processing Society of Japan(IPSJ) 44(4), 341–347 (2003) (in Japanese)
11. Ubayashi, N.: Modeling techniques for designing embedded software. Journal of Information Processing Society of Japan(IPSJ) 45(7), 682–692 (2004) (in japanese)
12. Nakamoto, Y., Takada, H., Tamaru, K.: Current state and trend in embedded systems. Journal of Information Processing Society of Japan(IPSJ) 38(10), 871–878 (1997) (in Japanese)
13. Nakashima, S.: Introduction to model-checking of embedded software. Journal of Information Processing Society of Japan(IPSJ) 45(7), 690–693 (2004) (in Japanese)
14. Ogasawara, H., Kojima, S.: Process improvement activities that put importance on stay power. Journal of Information Processing Society of Japan(IPSJ) 44(4), 334–340 (2003) (in Japanese)
15. Student: The probable error of a mean. Biometrika 6(1), 1–25 (1908)
16. Takagi, Y.: A case study of the success factor in large-scale software system development project. Journal of Information Processing Society of Japan(IPSJ) 44(4), 348–356 (2003) (in Japanese)
17. Tamaru, K.: Trends in software development platform for embedded systems. Journal of Information Processing Society of Japan(IPSJ) 45(7), 699–703 (2004) (in Japanese)
18. Watanabe, H.: Product line technology for software development. Journal of Information Processing Society of Japan(IPSJ) 45(7), 694–698 (2004) (in Japanese)
19. Welch, B.L.: The generalization of student's problem when several different population variances are involved. Biometrika 34(28) (1947)

A Study on Guiding Programmers' Code Navigation with a Graphical Code Recommender

Seonah Lee and Sungwon Kang

Abstract. While performing an evolution task, programmers spend significant time trying to understand a code base. To facilitate programmers' comprehension of code, researchers have developed software visualization tools. However, those tools have not predicted the information that programmers seek during their program comprehension activities. To responsively provide informative diagrams in a timely manner, we suggest a graphical code recommender and conduct an iterative Wizard of Oz study in order to examine when and what diagrammatic contents should appear in a graphical view to guide a programmer in exploring source code. We found that programmers positively evaluate a graphical code recommender that changes in response to their code navigation. They favored a graphical view that displays the source locations frequently visited by other programmers during the same task. They commented that the graphical code recommender helped in particular when they were uncertain about where to look while exploring the code base.

1 Introduction

When conducting software evolution tasks, programmers typically spend a considerable amount of time—more than 50% of their time—performing activities related to understanding an existing code base [9]. These activities include searching, navigating and debugging. Among these activities, programmers spend 35% of their time navigating code base [13]: programmers usually navigate several pieces of a code base, gather those pieces relevant to a required change, and build their understanding of the interrelationships between those pieces.

The main cause of the excessive time spent on understanding code base is that programmers have a difficulty in efficiently navigating the code base to find pieces of code that are relevant to a required change. Programmers still must randomly move around the code base through the structural relationships of several pieces of code. After reading a piece of code, programmers determine if the piece is relevant to the required change or not. The more often a programmer navigates

---

Seonah Lee · Sungwon Kang
Department of Computer Science, KAIST
e-mail: {saleese, sungwon.kang}@kaist.ac.kr

R. Lee (Ed.): Software Eng. Research, Management & Appl. 2011, SCI 377, pp. 61–75.
springerlink.com

non-relevant pieces of a code base, the more effort the programmer expends on reading and understanding those pieces [4].

For three decades, many software visualization tools have developed to facilitate programmers' comprehension of code [28]. However, programmers rarely use software visualization tools for their code comprehension [3]. Don't they want to use software visualization tools for their code comprehension? Apparently programmers want to use a software visualization tool for their code comprehension. A study conducted by Lee and colleagues [15] showed that programmers still expect to use software visualization tools for their code comprehension. The study also showed that 17 out of 19 participants desired to glance over a graphical view of the unfamiliar parts of code base. In particular, they suggested displaying the context of the code a programmer is currently looking at. We then notice that software visualization tools have not responsively provided the relevant information that programmers seek during their activities.

To facilitate a programmer's comprehension of code, a software visualization tool, we believe, is required to instantly recommend new parts of a code base which are relevant to a change (without requiring additional tuning efforts from the programmer.) To make a software visualization tool that timely provides informative diagrams, we need to know what diagrammatic contents programmers expect to see and when they want to see them. To find answer to the questions, we conduct a Wizard of Oz study. In an iterative Wizard of Oz study [8][11], we simulated a Graphical Code Recommender (GCR) that creates a graphical view on the fly as a programmer navigates the code base. In the Wizard of Oz study, programmers were given a graphical view while performing software evolution tasks. They were observed and interviewed, and their responses and feedback were gathered to discuss the diagrammatic contents and display timing in a graphical view.

This paper is organized as follows. Section II discusses the related work. Section III describes the Wizard of Oz study that we conducted. Section IV explains the findings through iterations during the study. Section V summarizes the participants' feedback and responses. Section VI discusses the positive and negative features of the given prototype of a graphical code recommender. Section VII explains possible threats to the validity of our research and Section VIII concludes the research.

2 Related Work

To our knowledge, this graphical code recommender is the first attempt that makes a code recommendation in software visualization tools in order to facilitate programmers' comprehension of code. We investigated several empirical studies on program comprehension as groundwork for our research (Section II.A). We also review the arguments in developing software visualization tools for programmers' comprehension of code (Section II.B). We also check if our work has a novelty in diagrammatic supports and makes a further step to resolve the struggling in this field (Section II.C). The emerged graphical code recommender mainly differs from other history-based approaches in using a graphical view (Section II.D).

2.1 Empirical Studies of Programmers

Elucidating how programmers understand source code during evolution tasks is a key to developing effective support for program comprehension. When we investigated the empirical studies of programmers [1][13][15][16][23][29], we were able to identify three programmers' practices. First, during program comprehension activities involved in an evolution task, programmers seek new parts of a code base that are relevant to a required change [13][16]. Second, while seeking new parts of a code base, programmers continually ask questions and conjecture answers [16][24]. Third, programmers as humans have a cognitive limitation. A human's working memory is limited in the number of chunks that can be processed at one time [1][29]. To meet those programmers' practices, our graphical code recommender is to help a programmer find new parts of a code base that are relevant to a required change. It is to provide a small amount of source information by degrees, which would be possibly an answer to the programmer's question during their code navigation.

2.2 Arguments on Diagrammatic Support for Program Comprehension

To facilitate programmers' comprehension of code, many software visualization tools have developed as comprehension aids [28]. However, little impact of those tools on program comprehension activities has been demonstrated [10]. Reiss [21] declared that software visualization tools have generally failed to impact on programmers' code comprehension. A recent empirical study conducted by Cherubini and colleagues in 2007 [3] found that programmers rarely use software visualization tools for their code comprehension. Despite this reality, there is a belief that a software visualization tool can improve program comprehension. Petre et al. [20] maintained that a software visualization tool would help, if it is an effective cognitive tool as much as it is supposed, and identified cognitive questions about software visualization. Storey et al. [29] identified cognitive design elements of a software visualization tool, based on program comprehension models and a programmer's cognitive overload. An empirical study conducted by Lee and colleagues in 2008 [15] found that programmers still expect to use software visualization tools for their program comprehension. For that, in the study, 17 of 19 participants desired to glance over a graphical view of the unfamiliar parts of code base where they explore. We notice that while program comprehension involves intensive information seeking activities [14], traditional software visualization tools have not provided the relevant information that programmers seek during their program comprehension activities [28]. We view this shortcoming as a main reason that programmers do not benefit from a software visualization tool.

2.3 Diagrammatic Support for Program Comprehension

Software visualization research for program comprehension has developed in two main directions. The first one provides a full overview of a software system. For example, Code Canvas [7] provides an entire map of source code and enables a programmer to use multiple canvases by filtering and customizing the entire map. However, it requires the programmer to tune up a visualized view. The second direction incrementally visualizes the parts of code base that a programmer navigates. For example, Code Bubble [2] enables a programmer to review multiple code fragments in parallel in a huge screen as navigating code base. This approach does not inform about the parts that a programmer has not navigated yet. Different from the two main directions, Our GCR is to predict and visualize the relevant information that programmers are seeking. As a history-based visualization tool, NavTracks [25] is similar to ours. However, NavTracks is limited in visualizing the files that a programmer has visited. GCR focuses on predicting in a timely manner the parts of code base that programmers have not explored but need to see.

2.4 History-Based Code Recommenders

The approaches of leveraging programmers' history have emerged. An approach, ROSE (Reengineering Of Software Evolution) [30], applies a data-mining technique to version histories to suggest further changes. Other approaches, TeamTracks [6] and Mylyn [12], gather the program elements that a programmer frequently visited to provide those program elements that a programmer may be interested in. However, TeamTracks and ROSE are limited to recommending those program elements co-visited or co-changed with a method that a programmer selects, most of those recommended ones were not relevant to the given task contexts. Mylyn does not automatically recommend program elements. Rather, it requires a programmer's explicit effort in identifying and selecting a task. Our GCR differs from other history-based approaches in a way of recommending program elements in visualizing the code recommendations in a graphical view.

3 Wizard of Oz Study

A Wizard of Oz study [8][11] is a research method where participants interact with an intelligent system, which they believe to be autonomous but is actually being operated by an unseen human being—a wizard operator. A wizard operator is a person who is behind the wall from a participant, and responded to the participant. The purpose of the Wizard of Oz study is to draw out the participants' responses to a working-in-progress system by providing the outputs of the system by hand.

We conducted a Wizard of Oz study, because it enabled us to gather the participants' responses in a to-be environment. In our study, we simulated a Graphical Code Recommender (GCR) that creates a graphical view on the fly as a programmer navigates the code base. By simulating a Graphical Code Recommender which instantly displays different diagrammatic contents in different time intervals, we could gather the participants' opinions on which diagrammatic contents and which time intervals they prefer. In our Wizard of Oz study, programmers performed evolution tasks. While they were performing evolution tasks, they were given a graphical view. Different from the original Wizard of Oz study, a wizard operator was standing behind a participant's back, because the wizard operator should observe the participant's interactions with the code base, and control the display of diagrams.

To let the wizard operator instantly display a diagram, we prepared a set of diagrams, based on the interaction traces where twelve programmers had performed the programming tasks described in Table 1: they conducted the four tasks on the JHotDraw 6.01b code base. The interaction traces helped us anticipate the parts and paths that our participants would navigate in the case that they perform the same tasks. As a result, the wizard operator was able to put a diagram on the second screen which the participant may need, based on where the participant was navigating.

After the study session, programmers provided their feedbacks to the graphical view. By incorporating participants' opinions into the design of GCR, we intend to develop a tool that satisfies programmers' expectations.

3.1 Participants

Eleven participants were recruited for the study. To be eligible for the study, participants were required to have Java programming experience using Eclipse. The participants had, on average, 4 (std. dev. 1.9) years of experience with Java, 2.6 (std. dev. 1.1) years with Eclipse, and 5 (std. dev. 4.9) years of industrial experience. The average age was 31.6 (std. dev. 4.3). All participants were male. We identified each participant with a two-character combination: the first represents the iteration number and the second the individual. The participants were four graduate students (1a, 1b, 3j, 3k), three industrial programmers from the same company (2c, 2d, 2e), and four industrial programmers from four different companies (3f, 3g, 3h, 3i).

3.2 Study Format

We conducted a Wizard of Oz study with three iterations. They are shown in Table 1. We followed the guideline of an iterative Wizard of Oz study [8] to elaborate a design. Through iterations, the role of a wizard operator was reduced from a controller to a moderator, and finally to a supervisor. The participants' responses from iteration were used to revise the prototype for the next iteration. In this way, GCR evolved from a paper prototype to a computerized prototype.

In the first iteration, we tested what contents should be shown in a graphical view between two different types of graphical contents (NI: Navigational Information and SI: Structural Information). Participants 1a and1b were asked to perform task 2-Arrows. While a participant was performing task 2-Arrows, a wizard operator observed the participant's interactions with code base, standing behind the

Table 1 Iterations of Our Wizard of Oz Study

Iteration	Prototype	Study format
1st iteration (controlled by a wizard operator)	A paper prototype consisted of two types: Structural Information (SI) and Navigational Information (NI). SI type consisted of diagrams that display 2~5 classes that have structural relationships. NI type consisted of diagrams that display classes with the methods frequently visited by other programmers in [6].	Two participants (1a and 1b) performed one evolution task (Task 2): Participant 1a with the SI type and participant 1b with the NI type. The wizard operator observed the participants' interactions with the code base, and displayed a diagram on a second screen.
2nd iteration (partly controlled by a wizard operator)	A computerized prototype displayed classes with the methods most frequently visited around the programmer's current location in the code. A simple rule of displaying a diagram was embedded.	Three participants (2c, 2d, and 2e) performed two evolution tasks (Tasks 2 and 3). If the participants gained no diagram for 10 minutes, the wizard operator manually loaded a diagram in a graphical view.
3rd iteration (not controlled by a wizard operator)	A revised prototype displayed diagrams in the shorter time interval than the previous version did, i.e. 1~3 minutes rather than 5~10 minutes.	Six participants (3f, 3g, 3h, 3i, 3j and 3k) performed three evolution tasks (Tasks 1, 2, and 3). The wizard operator observed the participants' interactions with the code base but did not control the graphical view.

participant and manually displaying a diagram. At the end of session, the participant was interviewed.

In the second and third iterations, we checked when the NI typed contents should be shown and what kind of programmers' interactions should be counted to show such contents. Participants were asked to perform tasks 1, 2, and 3. While a participant was performing those tasks, a computerized prototype automatically displayed diagrams. A wizard operator observed the participant's interactions with code base, and manually loaded a diagram only in the case that the participants were not given any diagrams.

3.3 Tasks

The participants conducted the tasks on JHotDraw 6.01b, which were designed during an earlier study [6]. The four tasks are described in Table 2.

Table 2 Tasks Used in the Earlier Study [22]

Task	Description
1-Nodes	Make a connection from and to the corners of a node
2-Arrows	Change an arrow tip according to the menu
3-Duplicate	Make a short-cut key for the duplicate command
4-Size	Inform the size of an active pane in the status bar

3.4 Prototype

The Graphical Code Recommender (GCR) displays the parts of the code base that the programmer has not explored but will likely need to explore, given the programmers' current navigation behavior.

We came up with a GCR as a tool that complies with programmers' practices as well as facilitates programmer's comprehension of code. Our prototype of GCR evolved through iterations. Initially we created paper prototypes to know what kind of contents programmers would like to have. Then, we made up a computerized prototype to know the time intervals they want to have such diagrammatic contents. The use of the prototype was to gather the participants' detailed responses under the particular situation where they performed program comprehension activities and were given diagrammatic contents at the same time.

The paper prototype consists of two different kinds of diagrammatic contents. The first kind presents the structural relationships among two to four classes. The structural relationships in the code became explicit depicted in the diagrams. The second kind presents the classes, methods and fields that the participants in an earlier study [22] frequently navigated during an evolution task.

The computerized prototype, which was used in the second and third iterations, recommends the most frequently visited ones as a participant navigates a code base, based on the interaction traces gathered in the study [22]. The computerized prototype is illustrated in Fig. 1. It monitors programmer's interactions with a code base, continuously tracking up to 15 program elements that the programmer has recently navigated. When a programmer has navigated the three program elements that were counted as frequently visited ones interaction traces gathered in the study [22], the prototype recommends the other program elements frequently navigated together with those three program elements. It intends to display a set of program elements most frequently navigated in the similar context to the current programmers' code navigation. The computerized prototype also supports a direct location of code base from a diagram. If a programmer clicks on a source location in a graphical view, the source location will be revealed in the editor, enabling the programmer to easily access to the source locations that were displayed in the graphical view.

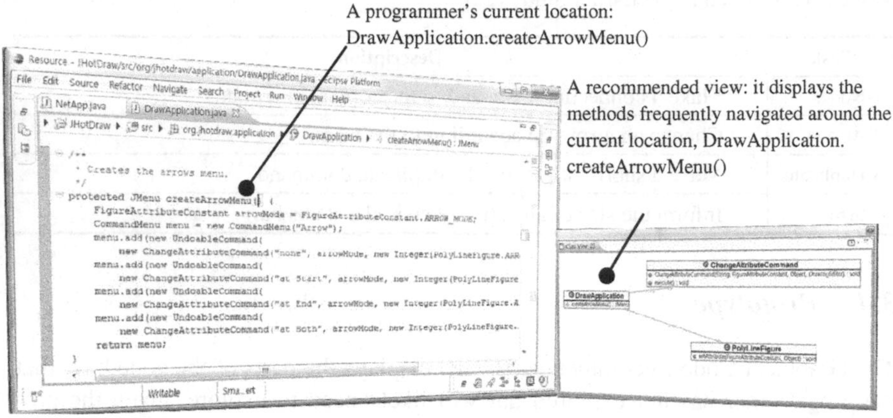
A programmer's current location:
DrawApplication.createArrowMenu()
A recommended view: it displays the methods frequently navigated around the current location, DrawApplication.createArrowMenu()

Fig. 1 The NavClus' process of clustering and recommending collections of code

Between the second and third iterations, we altered the time points to display diagrammatic contents in a shorter time interval and changed the way of capturing a programmer's interactions with code base.

3.5 Equipment

We set up the experimental environment in the participants' office for their time and convenience. We used a laptop computer with a 17-inch monitor, and a potable 9-inch side monitor. The Eclipse editor was placed on the main monitor. The graphical view was placed on the side monitor so as not to impede the participant's view of the editor. For data collection, we used a camera to take pictures of the displayed diagrams, and a microphone to record the participants' voices.

3.6 Procedure

Before the session, the participants were provided with a 4-page JHotDraw Tutorial and 1-page tool usage. They were allowed 10 minutes to become familiar with the experimental settings.

During the session, the participants were then given 1 hour to perform tasks of the previous study as described in Table 2. A wizard operator stood behind the participant's back, monitoring the participant's interactions with the code base, and sometimes placing a diagram in front of the participant. The wizard operator also observed the participants' behavior and took a picture of each diagram that the participants saw. Before beginning the task, we encouraged the participants to use a think aloud protocol, but did not remind them of using this protocol during the study.

After the session, the participants were then interviewed with prepared questions for 30 minutes. In the last iteration, a questionnaire was used, on which users evaluated the usefulness of the given views with the following scale: 5 – Excellent, 4 – Very Good, 3 – Good, 2 – Fair, 1 – Poor. We then interviewed the participants with the interview questions described in Section 3.6.

3.7 Questions

We interviewed the participants with three main questions: (i) whether programmers find a recommended view useful when navigating the code; (ii) what diagrammatic contents programmers expect to see in the view; and (iii) when programmers expect to see the diagrammatic contents.

4 Findings through Iterative Wizard of Oz Study

We conducted the first iteration to determine the more effective one of the two types of graphical content, Structural Information (SI) and Navigational Information (NI). According to the result, we provided NI typed information during the second iteration, and tested when such content should be shown on the screen. According to the result of the second iteration, we reduced the time interval of displaying diagrams. During the third iteration, we checked how the participants responded to the revised prototype.

4.1 First Iteration

Two participants (1a, 1b) responded positively but at different degrees. Participant 1a, who was given SI-typed diagrams, positively commented that the given diagrams helped him verify what he had already anticipated. Though, he said that he was not guided by the diagrams. Participant 1b, who was given NI-typed diagrams, stated that the given diagram was very good because he easily found where to look. He kept asking about how we were able to prepare such a valuable diagram. As participant 1b responded much more positively than 1a did, we conjectured that NI-typed diagrams were most likely to be capable of guiding programmers.

4.2 Second Iteration

Two of three participants (2c, 2e) responded positively, while one participant (2d) negatively. Participant 2e gave a reason similar to the comment made by participant 1b. Participant 2e commented, "It helped me find entry points [...] I was able to select a method to examine from here." In contrast, participant 2d commented, "I did not understand how this class view showed up, and what rules made it appear."

It was difficult to identify an optimal time interval with a simple rule. Participants 2c and 2d were given diagrams every 3~5 minutes; 2c commented positively, while 2d responded negatively. Participant 2e was given a diagram by a wizard operator after 10 minutes. He greatly appreciated it commenting that the diagram helped when he was uncertain about where to look.

4.3 Third Iteration

Three of six participants (3g, 3h, 3k) evaluated the prototype as "Very Good," two participants (3i, 3j) as "Good," and one participant (3f) as "Fair." Participant 3h reported, "It provides a crucial hint." Participant 3k was overheard saying, "Uh, here are all the answers." We interpret this result as that it helped a programmer navigate toward task relevant source locations.

Participants 3f, 3g, 3h, 3i, 3j and 3k were given diagrams in 1~3 minutes. They stated that they were not disturbed by the graphical view. They only referred to the graphical view when they wanted to find a clue in accomplishing an evolution task. They skipped some diagrams, while being absorbed in reading source code.

5 Study Results

We summarize the participants' responses to our questions to suggest a graphical code recommender that is based on their responses.

5.1 Do Programmers Find a Graphical View Useful?

Many participants (9 of 11: 1a, 1b, 2c, 2e, 3g, 3h, 3i, 3j, 3k) positively evaluated the given graphical view, while a few participants (2 of 11: 2d, 3f) negatively evaluated. The first reason that most of the participants favored the prototype was that they could find the classes, methods, and fields relevant to the task that they were performing. In particular, when we use a questionnaire in the third iteration, the six participants (3f, 3g, 3h, 3i, 3j, 3k) rated as shown in Fig. 2.

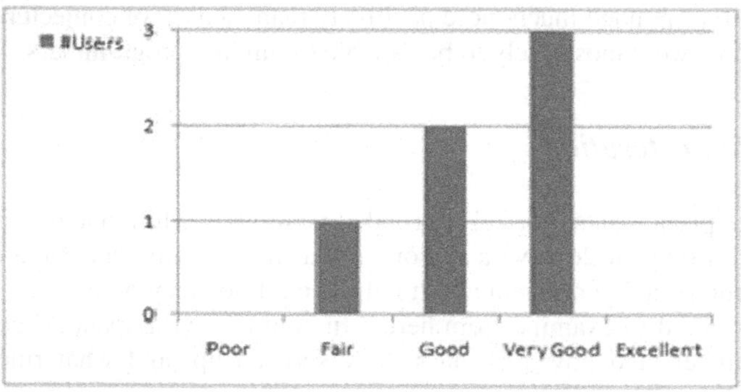

Fig. 2 The Usefulness of the Graphical Code Recommender

Table 3 The Diagrammatic Contents that the Participants Expected to See

Diagrammatic Contents	Participant ID
Task relevant methods	2c, 2e, 3g, 3h, 3k
Relationships among classes	1a, 1b, 2c
Description of methods	2e, 3i
A class hierarchy	2d, 3f, 3j
An overview	2c, 2d, 3k

5.2 *What Diagrammatic Contents Do Programmers Find Useful in a Graphical View?*

Participants found the diagrammatic contents listed in Table 3 useful. First, the participants found task relevant methods and relationships among classes useful. Five participants (2c, 2e, 3g, 3h, 3k) appreciated task relevant methods. They commented that the given graphical views helped them by narrowing the scope of examination while they were looking for several pieces of source code relevant to a given task. Participant 2c mentioned, "Here in createArrowMenu(), I found where to look to solve this problem. It reduces the candidate scope of examination." Participant 3h told, "The fields actually help to find the code to change. The relationships indirectly help verify how two classes are connected with each other." Three participants (1a, 1b, 2c) prized explicit relationships among classes. Participant 1a commented that he usually looked at the relationship between two classes but not their methods. Participant 1b noted, "The most useful information is the relationship between DrawApplication and ChangeAttributeCommand in the first diagram" (see Fig. 1).

As diagrammatic content to display, the participants desired to see more abstract information as well as more detailed information. Three participants (2d, 3f, 3j) mentioned a class hierarchy. Participant 2d said that for a better overview it should show a class hierarchy that includes the classes that were already navigated. Two participants (2c, 2d) expected to have a more abstract view than the displayed graphical view. Participant 2c commented that such a view would assist a team leader who frequently cases around unfamiliar code base. For that, he anticipated, a more abstract view would be needed. In contrast, participant 3k discovered that the given view had displayed an overview of the parts of source code where he had navigated and modified.

5.3 *When Do Programmers Find a Graphical View Useful?*

In overall, the time intervals that provide diagrammatic contents to the participants have been reduced from 5~10 minutes to 1~3 minutes through iterations, as shown in Table 4. However, it is difficult to identify an optimal time interval with a simple rule, because the time point when a programmer needs a diagram significantly

Table 4 The Intervals that Provide Diagrammatic Contents to the Participants

Time Interval	Participant ID
5~10 minutes	1b, 2e
3~5 minutes	1a, 2c, 2d
1~3 minutes	3f, 3g, 3h, 3i, 3j, 3k

relies on the programmer's individual circumstances. Participants 1b and 2e, who were given a graphical view in 5~10 minutes interval, strongly appreciated it because the given graphical view helped when they were uncertain about where to look. Participants 2c and 2d, who were given a different graphical view in 3~5 minutes, with a simple rule embedded in a computerized prototype, expressed contrasting opinions on a graphical view. Participants 3f, 3g, 3h, 3i, 3j and 3k were given a different but similar graphical view in 1~3 minutes. They only referred to the graphical view when they wanted to find a clue in accomplishing an evolution task. They skipped some graphical views, because they were so absorbed in reading source code. They did not feel that they were disturbed by a graphical view that changed every 3 minutes. As long as a graphical view brought in new and useful information that would help a programmer's understanding of code to perform an evolution task, the programmers appreciated it.

6 Discussion

Though our aim was to explore desired diagrammatic content and optimal time points to provide such content, we also received many feedbacks on the current version of the prototype.

The participants positively evaluated the prototype which recommended the classes, methods, and fields relevant to the task that they performed, more specifically had frequently visited in the same task context. One of the participants (3h) clearly addressed that the graphical elements were the more important information than graphical relations in a graphical view. It belied the traditional claim that a graphical view would be useful than a textual view because of explicit representation of relationships. The support of the prototype for the direct linking from graphical elements to the code base also received the participants' appreciations. It was intended to save the participants' effort in searching the program elements that they found interesting in a graphical view. Some of the participants (3h, 3i, 3j, 3k) actively used the capability to access the program elements that were interesting in a graphical view. Other participants also positively mentioned this capability in their interview. It was interesting that none of the participants was bothered by the graphical views in the side monitor.

Through this study, we also found the improvement points of the given prototype. Many of the participants expected to see a broader scope of an overview than the current version of a prototype. They expressed their expectation in different terms, such as a class hierarchy, a more abstract view, etc. If we pursue a hybrid approach of combining historical analysis and structural analysis, we could provide the expanded scope of recommendations in a graphical view. Another comment from the participants was about the recommendation rules of a graphical view. They commented that the recommendation rule of displaying graphical contents were not intuitive. This was one of the reasons that participant 2d negatively evaluated the prototype. We are considering changing the prototype to instantly display the program elements that a programmer selects even before making a recommendation. We expect that displaying those element counted for a recommendation will enable the participants to anticipate when the recommendation will occur.

7 Threats to Validity

The results of our study could be affected by different graphical views being given at different time intervals and the wizard operator's subjectivity. In spite of these dependencies, we were able to design a Graphical Code Recommender by conducting the iterative Wizard of Oz study and exploring design parameters—such as the content of a graphical view and the timing of its presentation—that can effectively guide a programmer's code navigation.

8 Conclusion and Future Work

Our Wizard of Oz study reveals that programmers favored a graphical view that displayed the source location frequently visited by other programmers who perform the same task. The participants stated that the graphical view was very helpful, in particular when they were uncertain about where to look while exploring the code base. We interpret this result as an evidence that a graphical code recommender that instantly recommends new parts of a code base that are pertinent to a programmer's task would significantly facilitate programmers' comprehension of code.

For further improvement, we plan to investigate about the effective amount of information that a graphical view displays at a time and to develop intuitive recommendation rules that allow programmers to predict when a recommendation happens.

Acknowledgments. We thank Gail C. Murphy for her advice and the participants of the survey for their comments. This research was supported by Basic Science Research Program through the National Research Foundation of Korea (NRF) funded by the Ministry of Education, Science and Technology (20100022431).

References

[1] Baddeley, A.: Working Memory: Looking Back and Looking Forward. Neuroscience 4(10), 829–839 (2003)

[2] Bragdon, A., Zeleznik, R., Reiss, S.P., Karumuri, S., Cheung, W., Kaplan, J., Coleman, C., Adeputra, F., LaViola Jr., J.J.: Code bubbles: a working set-based interface for code understanding and maintenance. In: Proc. CHI 2010, pp. 2503–2512. ACM, New York (2010)

[3] Cherubini, M., Venolia, G., DeLine, R., Ko, A.J.: Let's Go to the Whiteboard: How and Why Software Developers Use Drawings. In: Proc. CHI 2007, pp. 557–566 (2007)
[4] Coblenz, M.J.: JASPER: facilitating software maintenance activities with explicit task representations, TR CMU-CS-06-150, School of Computer Science, Carnegie Mellon University, Pittsburgh, PA (2006)
[5] Cox, A., Fisher, M., Muzzerall, J.: User Perspectives on a Visual Aid to Program Comprehension. In: Int'l Workshop on Visualizing Software for Understanding and Analysis(VISSOFT), pp. 70–75 (2005)
[6] DeLine, R., Czerwinski, M., Robertson, G.: Easing Program Comprehension by Sharing Navigation Data. In: Proc. VLHCC 2005, pp. 241–248 (2005)
[7] DeLine, R., Venolia, G., Rowan, K.: Software development with code maps. ACM Queue 8(7), 10–19
[8] Dow, S., MacIntyre, B., Lee, J., Oezbek, C., Bolter, J.D., Gandy, M.: Wizard of Oz Support throughout an Iterative Design Process. IEEE Pervasive Computing, 18–26 (2005)
[9] Fjeldstad, R., Hamlen, W.: Application program maintenance-report to our respondents. Tutorial on Software Maintenance, 13–27 (1983)
[10] Hendrix, T., Cross II, J., Maghsoodloo, S., McKinney, M.: Do Visualizations Improve Program Comprehensibility? Experiments with Control Structure Diagrams for Java. In: 31st SIGCSE Technical Symp. on Computer Science Education, pp. 382–386 (2000)
[11] Kelley, J.F.: An Iterative Design Methodology for User-friendly Natural Language Office Information Applications. ACM TOIS 2(1), 26–41 (1984)
[12] Kersten, M., Murphy, G.C.: Using task context to improve programmer productivity. In: Proceedings of the 14th ACM SIGSOFT International Symposium on Foundations of Software Engineering, Portland, Oregon, USA, November 5-11 (2006)
[13] Ko, A.J., Aung, H., Myers, B.A.: Eliciting Design Requirements for Maintenance-Oriented IDEs: A Detailed Study of Corrective and Perfective Maintenance Tasks. In: Proc. ICSE 2005, pp. 126–135 (2005)
[14] Ko, A.J., Myers, B.A., Coblenz, M.J., Aung, H.: An Exploratory Study of How Developers Seek, Relate, and Collect Relevant Information during Software Maintenance Tasks. IEEE TSE 32(12), 971–987 (2006)
[15] Lee, S., Murphy, G.C., Fritz, T., Allen, M.: How Can Diagramming Tools Help Support Programming Activities. In: Proc. VLHCC 2008, pp. 246–249 (2008)
[16] Letovsky, S.: Cognitive Processes in Program Comprehension. In: Proc. ESP 1986, pp. 58–79 (1986)
[17] Mayrhauser, A., Vans, A.M.: Comprehension "Processes during Large Scale Maintenance. In: 16th Int'l Conf. on Software Eng., pp. 39–48 (1994)
[18] Miller, G.A.: The magical number seven plus or minus two: some limits on our capacity for processing information. Psychological Review 63(2), 81–97 (1956)
[19] Pennington, N.: Stimulus Structures and Mental Representations in Expert Comprehension of Computer Programs. Cognitive Psychology, 295–341 (1987)
[20] Petre, M., Blackwell, A.F., Green, T.R.G.: Cognitive Questions in Software Visualization. In: Software Visualization: Programming as a Multi-Media Experience. MIT Press, Cambridge (1997)
[21] Reiss, S.: The Paradox of Software Visualization. In: Int'l Workshop on Visualizing Software for Understanding and Analysis (VISSOFT), pp. 59–63 (2005)

[22] Safer, I.: Comparing Episodic and Semantic Interfaces for Task Boundary Identification, M.Sc. Thesis (2007)
[23] Shaft, T.M., Vessey, I.: The Relevance of Application Domain Knowledge: Characterizing the Computer Program Comprehension Process. JMIS 15(1), 51–78 (1998)
[24] Sillito, G.C., Murphy, G.C., De Volder, K.: Questions programmers ask during software evolution tasks. In: Proceedings of the 14th ACM SIGSOFT International Symposium on Foundations of Software Engineering, Portland, Oregon, USA, November 5-11 (2006)
[25] Singer, R., Elves, Storey, M.-A.D.: NavTracks: Supporting Navigation in Software Maintenance. In: Proc. ICSM 2005, pp. 325–334 (2005)
[26] Sim, S.E., Holt, R.C.: The Ramp-Up Problem in Software Projects: A Case Study of How Software Immigrants Naturalize. In: 20th Int'l Conf. on Software Eng. (ICSE), pp. 361–370 (April 1998)
[27] Sinha, V., Karger, D., Miller, R.: Relo: Helping Users Manage Context during Interactive Exploratory Visualization of Large Codebases. In: Proc. VLHCC 2006, pp. 187–194 (2006)
[28] Storey, M.-A.D.: Theories, Methods and Tools in Program Comprehension: Past, Present and Future. In: Proc. ICPC 2005, pp. 181–191 (2005)
[29] Storey, M.-A.D., Fracchia, F., Müllecr, H.: Cognitive Design Elements to Support the Construction of a Mental Model during Software Visualization. In: Proc. IWPC 1997, pp. 17–28 (1997)
[30] Zimmermann, T., Weisgerber, P., Diehl, S., Zeller, A.: Mining Version Histories to Guide Software Changes. In: Proc. ICSE 2004, pp. 563–572 (2004)

A Testing Frameworks for Mobile Embedded Systems Using MDA

Haeng-Kon Kim and Roger Y. Lee

Abstract. Embedded system can give you many benefits in putting it in your device, such as mobile phones, appliances at home, machines at the bank, lottery machine and many more, just make sure it is undergoing in embedded systems testing to have the device check. You must know that putting an embedded system in any of your device (either at home or in your business) can vary be helpful in your daily life and for the near future. One of the important phases in the life cycle of embedded software development process is the designing phase. There are different models used in this particular phase including class diagrams, state diagrams and use cases etc. To test the conformance of the software it is very essential that test cases should be derived from these specific models. Similarly regressions testing through these models are very significant for testing of modified software. There are several regression testing approaches based on these model in literature. This survey report is the analysis of the model based regression testing techniques according to the parameter identified during this study. The summary as well as the analysis of the approaches is discussed in this survey report. In the end we concluded the survey by identifying the areas of further research in the field of model based regression testing.

Keywords: MDA, Embedded testing, Regression testing, model based regression testing, UML regression testing, Testing evaluation parameters.

1 Introduction

To inject quality into the software, testing is important activity in the software development process. Testing can be done in several ways, model based, code based

Haeng-Kon Kim
Department of Computer Engineering, Catholic University of Daegu, Korea
e-mail: hangkon@cu.ac.kr

Roger Y. Lee
Software Engineering & Information Technology Institute, Central Michigan University, USA
e-mail: leelry@cmich.edu

R. Lee (Ed.): Software Eng. Research, Management & Appl. 2011, SCI 377, pp. 77–94.
springerlink.com

and specification based etc. Many model based testing approaches are developed for the purpose of quality improvements of the software. In model based testing [1], the test cases are derived from a certain model (class diagrams, state diagrams, activity and use cases etc).

Since a model of software represents its requirements, the derivation of test cases from that model is to determine the quality of the software towards its conformance with requirements. Regression testing is an important part of software development life cycle since it revalidates the modified software [2]. The goal of the regression testing is to verify changes and to show that the changes do not harmfully affect other parts of the software. (Overview of the regression testing is discussed in section 2).

With the existence of model based testing, the existence of the model based regression testing is essential. For state of the art regression testing, the analysis of the existing model based regression testing approaches [3,4,5] is necessary, which is the purpose of this paper.

These approaches as a whole identify the differences between the original version of the software and the changed version of the software on different basis. The basis of the different identification between the original and the modified versions normally depends upon the characteristics of the models used for the regression test generation. Every model has their own characteristics and the testers try to identify the changes according to these characteristics. To avoid the ambiguities we explain the word characteristic of any model with respect to the example. In a class diagram (for example) the word characteristics refer to the attributes and operations [2]. Here we consider attributes and the operations as characteristics of a class diagram.

After the change is identified between both the old and the new version of the software, it has to be tested, whether the newly changed part is working properly and the unchanged portion of the software is not affected with this specific change.

In the next section we present background about the regression testing, in section 3 the survey of existing model based techniques is presented with classification and analysis. Model based regression testing approaches are evaluated in section 4, finally we concluded in section 5.

2 Related Research

2.1 Type of Automated Testing

Automated testing can be split into two different types: White box testing and Black box testing. These are common terms in the testing world. Black box testing is when a tester tests the software in an external way, usually knowing little to nothing about the internal workings of the program. White box testing is when a tester tests the software from the inside and knowledge of internal works is required. In the automated testing world, black box testing is GUI testing.

Black box testing has been slower to move towards automated testing compared to other types of testing, but has been a driving force in commercial testing software. This type of testing is slightly harder to automate because the tester is testing the program in a completely external way. Test scripts for back. With these playback scripts the tester can then set ending conditions that the testing environment can check to see if the test passes or fails. These environments work well testing executable GUIs.

The other type of black box automated testing is test scripts. These come less in the form of testing environments and more in the form of testing APIs. They give the tester a library of commands they can call to manipulate the screen and asserts to test results. Asserts are the core of the testing libraries and can take many forms. They all have the same general purpose, they take two parameters and perform a comparison of some type on them. They often have a logging ability also. These testing libraries allow a programmer to write a test case that can easily be run in batch scripts. This type of automated black box testing works well with web applications.

White box testing, or module testing, in a windows executable program situation is very straight forward. Most automated test helpers for this type of environment come in the form of an API containing assert methods. The tester can choose to test the system method by method, subsystem by subsystem, or any level of abstraction in between. These tests are usually written in the same programming language as the program being tested, because the tests interact directly with pieces of the program. This type of testing is best used for regression testing. Regression testing is where a tester runs a suite of tests, that they know worked on an old build of the software, and run it on a new build. This make sure that no functionality is lost between builds. As testers write test cases for each part of the system they can add that test script to the test suite. This test suite is where automated testing really shows its value. A group leader can run the test suite whenever their developers are not using their computers and in this way can continue to utilize the companies resources even after hours. Run the test suite when everyone leaves, when everyone comes back to work in the morning there are test results waiting from the previous night.

Embedded systems can insert some complexity into automated testing. With most embedded systems projects the tester must test their code on the target hardware. This often means that the testers automated test scripts can not be stored or run on the same machine as the code, like they can do with standard white box testing. This is where a more heavy duty testing API and/or environment is needed.

In order to white box test embedded systems the testers need a testing environment that can interface with the IDE/debugger that the developers used to code the system. The testers test scripts will not directly interact with the system, but instead interact with the debugger. The test scripts must be able to tell the debugger to set variables, read variables, set breakpoints, run to breakpoints, clear breakpoints, and many other actions that a tester would perform in a manual test.

This type of automated testing is expensive to do. The testers often have to write a testing environment/API from scratch to perform the functions that they require. This is an expensive and time consuming task. For small projects this might not be worth the cost. Yet with larger projects it is still worth the time and money. The time the project saves during regression testing and other testing phases in the long run makes up for the initial time delay.

2.2 Regression Testing

Testing can be used to build the confidence in the correctness of the software and to increase the software reliability. The major difference between the regression testing and the development testing is that during regression testing an established set of tests may be available for reuse. Regression testing is essential when software product has been changed and it is applied during maintenance. Regression testing ensures that the modified program meets its specification and new errors are uncovered [3].

Regression testing is expensive maintenance process which is responsible for revalidating the modified software. Maintenance cost of the software is high as compared to its development cost.. According to the new assumptions, the maintenance cost of the software is exceeded to 70.

Maintenance has three types; Perfective, adaptive and corrective [2]. Corrective maintenance is performed to correct error that has been uncovered in some part of the software. Adaptive maintenance is performed when software is modified to ensure its compatibility with the new environment in which it will operate. Perfective maintenance is performed to add new features to the software or to improve performance of the software. Regression testing is done in all these types [1].

There are two types of regression testing; Progressive regression testing and corrective regression testing. Corrective regression testing is applied when specification is not changed; probably some other change has been made i.e. correcting an error. In this case test cases can be reused. Progressive regression testing is applied when specifications have been changed and new test cases must be designed at least for the added part of the specification [1,2].

Regression testing uses two approaches to test the modified software; Retest all approach and selective test approach. Retest all approach chooses all test cases from the unchanged software to test the changed software, but the approach is time consuming as well as resource consuming approach. Selective retest approach chooses a subset of the tests from the old test suit to test the modified software [2]. In regression, testing selecting a suitable subset with an efficient algorithm is a major area of research.

2.3 Embedded System

Embedded systems can be roughly defined as a system that is not primarily a computer but contains a processor. But rather than focusing on a definition, it is useful to consider aspects that most embedded systems share, at least to some degree. Many embedded systems such as PDAs or cell-phones are high-volume, low-cost and low-margin. This requires use of the cheapest components possible, which typically means simple processors and small memory (RAM and NVRAM/flash). This causes embedded systems software to trade off maintainability aspects such as portability, clarity, or modularity for performance optimization aspects such as a small boot image footprint, a small RAM footprint, and small cycle requirements. The increased up-front software development costs and periodic maintenance costs are amortized by the high-volume sales, and outweighed by the continuous hardware cost savings of cheaper components. Many other embedded systems, though not so price-sensitive, have physical constraints on form factor or weight to use the smallest components possible. Again, this favors performance optimization at the cost of maintainability.

In addition to trading off portability, clarity, or modularity, embedded systems may also require optimization by using a low-level language, e.g. assembly rather than C, or C rather than code automatically generated from a UML model. However, this hand tuning is typically only applied to small portions of the software identified by the 90/10 guideline as being the major performance bottlenecks.

- **Embedded systems often have power limitations**

Many embedded systems run from a battery, either continually or during emergencies. Therefore, power consumption performance is favored in many embedded systems at the cost of complexity and maintainability.

- **Embedded systems are frequently real-time**

By nature, most embedded systems are built to react in real-time to data flowing to and through the system. The real-time constraints again favor performance aspects (particularly cycles usage) over maintainability aspects. There are generally both hard real-time constraints, which require an event to be handled by a fixed time, and soft real-time constraints, which set limits both on the average event response time and the permissible magnitude of outliers. Real-time operating systems use preemptive prioritized scheduling to help ensure that real-time deadlines are met, but careful thought is required to divide processing into execution contexts (threads), set the relative priorities of the execution contexts, and manage control/data flow between the contexts.

- **Embedded systems frequently use custom hardware**

Embedded systems are frequently comprised of off-the-shelf processors combined with off-the-shelf peripherals. Even though the components may be standard, the custom mixing and matching requires a high degree of cohesion between the hardware and the software – a significant portion of the software for an embedded system is operating system and device driver software. Though this low-level software is often available for purchase, license, or free use, frequently a large portion of the operating system for an embedded system is custom-developed in-house, either to precisely match the hardware system at hand, or to glue together off-the-shelf software in a custom configuration.

Often the functionality of an embedded system is distributed between multiple peer processors and/or a hierarchy of master/slave processors. Careful thought is required regarding the distribution of processing tasks across processors, and the extent, method, and timing of communication between processors.

Furthermore, many embedded systems make use of specialized FPGAs or ASICs, and thus require low-level software to interact with the custom hardware.

- **Embedded systems are predominantly hidden from view**

By nature, embedded systems typically have a limited interface with their user (real user or another component of the super-system). Thus, much of the system is developed to meet the software functional specifications developed during architecture and high-level design, rather than the user requirements.

- **Embedded systems frequently have monolithic functionality**

Most embedded systems are built for a single primary purpose. They can be decomposed into components, and potentially the components could have low cross-cohesion and cross-coupling. That is, each component could serve a distinct purpose, and the interactions between components could be restricted to a few well-defined points. Nevertheless, the system as a whole will not function unless most or all of the components are operational. A system that requires all components to function before the system as a whole achieves useful functionality is a "monolithic system". This non-linear jump in system functionality as a function of component functionality is in contrast to some other types of software, where the system may be 50.

For example, a space probe is built to travel by or to other planets and send back information about them. Though there are many lower-level responsibilities of the space probe components, such as targeting, landing, deploying sensors, deploying solar panels, and communications, each of these lower-level responsibilities is an indispensable component of the over-arching functionality. The space probe will be useless if any of these vital components is missing, even if all other components are completely functional.

Another example is a cell phone, in which all the sub-features such as the user interface, the cellular base station selection, the vocoder, and the communications protocols are all vital aspects of the over-arching goal to transfer bi-directional audio information between the user and specific remote nodes. These are in contrast to other software regimes, such as web services or desktop tools, in which lower-level responsibilities are more likely to contribute independently to the aggregate system functionality rather than serving as indispensable parts of a monolithic whole. Though the software components of an embedded system are combined into a monolithic functionality, the components themselves are often very distinct. Embedded systems will frequently combine software components that perform signal processing, low-level device driver I/O, communications protocols, guidance and control, and user interfaces. Each of these specialized components requires a distinct developer skill set.

- **Embedded systems frequently have limited development tools**

Though some software regimes have a whole host of tools to assist with software development, embedded systems software development are more limited, and frequently use only basic compiler tools. This is in part because embedded systems often use custom hardware, which may not have tool support, and because embedded systems are often real-time and performance constrained, making it difficult to freeze the entire execution context under the control of a debugger or transfer control and data between the embedded target and a host-based tool, or capture extensive execution-tracing logs.

Because of the limited choices of commercial tools for embedded systems software development, many embedded systems projects create their own tools to use for debugging and testing, or at least augment commercial tools with in-house tools.

- **Embedded systems frequently have stringent robustness requirements**

Embedded systems are often used in harsh environments and for mission-critical or medical purposes. Therefore, requirements for reliability, correct exception handling, and mean time between failures are typically more stringent for embedded systems than for many other types of software. This translates into rigorous development processes and testing requirements. In turn, this increases the overhead needed to make a release of software.

Some types of embedded systems are subject to regulatory requirements that purport to reduce fault rates by mandating the software development process, or at least specifying what documentation must accompany the embedded systems product. Furthermore, for several types of embedded systems, it is difficult or even impossible to upgrade firmware, which emphasizes the need to get it right in the systems initial commercial release.

- **Embedded systems are frequently very long-lived**

Embedded systems often stay in use for many years. Frequently the duration of support for an embedded system is far greater than the turnover rate of the original software developers. This makes it paramount to have good documentation to explain the embedded systems software, particularly since the source code itself may have its self-documentation quality compromised due to performance trade-offs.

3 Embedded Testing Using MDA

This paper presents an analysis of model based regression testing techniques with classification. These techniques generate regression tests using different system models. Most of the techniques are based on the UML models. The techniques in this survey use some models like, class diagrams, state machines diagrams, activity diagram, and use case diagrams etc. Most of the techniques use different types of state machines like state machine diagram, finite state machine and extended state machine diagrams. We classify the techniques presented in this paper according to kind of model the techniques are based on. Four classes are listed below:

1) State machine based approaches
2) Activity diagram based approaches
3) Model checking based approaches
4) Hybrid approaches

An evaluation of the techniques based on parameters (Identified during this survey study) is presented in section 4. Some of the parameters which we used are identified by Rothermel and Horrald [14]. Before a tabular evaluation of the model based regression testing approaches, all the techniques are classified into the unmentioned classes, briefly discussed and analyzed.

3.1 State Machine Based Approaches

The approaches classified as state machine based approaches use different types of state machines to generate regression test. We are going to present these approaches in a brief way.

(1) Embedded Test Suite Generation Using MDA

The model based approach given by Chen et al [6] is based on the extended finite state machine (EFSM). This approach is based on a 9-tuple attributes of EFSM. One of them is transition and the whole idea of regression testing lies in it (transitions of the EFSM consisting of 6-tuple). For the purpose of the regression testing they identified the dependences between these transitions. They concluded two types of dependences; Data dependence (DD) and the control dependences (CD). The data dependency is the usage of a variable by one transition that is set by another transition. While in control dependence one transition can effect the traversal of the other transition [6].

The basic structure of the EFSM is based on transitions and the data dependency is identified between the transitions. If one transition defines the data and the other transition defines the use of the data then data dependency exists between these two transitions. Basically the idea is to identify the def-use associations between the transitions as def-use associations are identified in data flow analysis. The test suite reduction is based on data dependency and the control dependency.

Control dependency also exists between two transitions, it identifies the traversal of a certain transition depends on another transition. To find out such dependencies the concept of the post domination is used. A state can post dominate another state of the any other transition if that state or transition has to traverse that certain state to reach the exit state. Control dependency exists if the transition is post dominated and the state of that transition is not post dominated.

Above mentioned dependencies are identified in three aspects, when EFSM model is modified; effect of the model on the modifications effect of the modifications on the model, and side effects to unchanged parts.

According to [6] there are three ways in which a model is considered as modified; addition of a transition, deletion of transition and change of transition. From these modifications all control dependencies and data dependencies are identified and further regression test suite is constructed on the basis of these dependencies.

The technique is tool supported and provides high efficiency and good coverage criteria. There is no case study found in this technique. Furthermore this technique is yet to be evaluated on an industrial based case study.

(2) Embedded Regression Test Reduction Using Dependence Analysis

The next technique discussed here is not purely a regression test selection technique. Basically the aim of the technique is to reduce the regression tests already selected for an EFSM model. This technique is based on the technique of regression test selection using dependence analysis by Chen and Probert [6]. In this approach [7] the test cases are generated using dependence analysis as this analysis was done in [6]. Two kinds of dependencies; DD and CD (discussed in previous technique) were identified and from these dependencies the regression tests were generated. Two kinds of dependencies i.e. data dependency and control dependency are discussed here briefly.

After the modifications are made the impact of the modifications is analyzed with respect to these two kinds of dependencies. By the analysis of the elementary modifications, three interaction patterns are calculated named as affecting interaction patterns, affected interaction pattern and side-effect interaction pattern then these interaction patterns are used to reduce the test suite [7].

This approach is basically a refinement of the regression test suite. So it is much helpful in the cost reduction of the regression testing. It is cost effective, efficient and importantly tool supported. It is yet to be evaluated on a larger case study to analyze its effectiveness and efficiency more deeply.

(3) Regression Test Suite Reduction Using Extended Dependence Analysis

This technique is highly based on the [6] and [7]. There is a little modification made in the approach discussed in [6]. The elementary modifications (EM) considered in [6] are addition of transition in EFSM model and the deletion of the transition in EFSM model. Now the extension is made in the approach to the reduction of regression test suite. All three types of EMs for the EFSM model [7] are considered in the reduction of the regression test suite. Third elementary modification i.e. the change of the transition is also involved and the data dependency and control dependency analysis as well [8]. In the earlier techniques the data and control dependencies were not analyzed.

Again in this approach three interaction patterns are computed and on the basis of these patterns the test suite is reduced as it is done in [7].

The major contribution of the approach is the consideration of the change of transition of the EFSM model as EM reduces the regression test suite. Previous technique [7] was not working with this type of modification so this approach removed the flaw of that technique.

One of the major plus points of this technique is its automation; no case study is presented and also not validated on an industrial based case study.

(4) Integrating White Box and Black Box technique for Class Level Testing

This is an approach for the regression testing based on the finite state machine. In this approach the researchers has proposed a class level regression testing technique by using both black box and white box testing techniques. Many techniques have been proposed for class level regression testing but most of these techniques focus on either black box testing or white box testing technique. Beydeda and Gruhu [12] have proposed a new approach at class level by integrating both white box and black box..An integrated approach can improve the efficiency of testing process in terms of cost and time.

Comparison is shown by using representation of a class called class control flow graph (CCFG). The comparison is carried out simultaneously by traversing both CCFG and comparing the nodes of the graph. In the second building block, the idea is to identify the definition and uses of each attribute and associate to other according to some data flow criteria. After association with each other test cases covering these def-use pairs are generated. This technique generates test cases by using class state machine (CSM) graph. The proposed method combines these techniques. Integrated white and black box testing technique operates on a graph called class specification implementation graph (CSIG) [12].

3.2 Activity Diagram Based Approaches

The approaches classified as activity diagram based approaches are used to generate regression tests through activity diagram models. Brief description and analysis of these approaches is presented in this section.

(1) Specification Based Regression Test Selection with Risk analysis

We consider this approach as a risk-based regression testing. In this approach the authors have considered the risks related to the software potential defects as a threat to the failure after the change as a significant factor, so a risk model is presented as well as the model of regression testing. Two type of test are to be included in the regression test suite, targeted tests and safety tests [9].

Targeted tests are the tests included in the regression testing for the reason of the change in the software while the safety test are those tests which are included into the regression tests on the basis of risk analysis. Combination of both targeted and safety tests are ultimately considered as regression tests [9].

To generate the targeted tests the activity diagram model is used. A way to generate test cases as well as regression tests from the activity diagram is presented. In the first step the regression test cases are generated by the change analysis in the activity diagram.

The second and important step is to generate test cases which are risk based. The author presented an approach for this purpose consisting of four steps. In the first step the cost of every test case is assessed. The cost of every test case is ranked through 1-5, where the lowest value depicts the lower cost and the high value as higher cost. The cost is measured in the form of consequences of the fault; the second step is to calculate the severity probability for each test case. It is calculated by multiplying the number of defects and the average severity. The severity probability is obtained by counting the uncovered defects by a certain test case. After the calculation of severity probability the risk exposure is calculated by the multiplication of then cost and severity probability of the defect. The value obtained is considered as the risk of the test case. Higher the value higher the risk is, and vise versa. Next the test cases with higher value of risk are chosen and included in the regression test suite. These are the safety tests which are run along with targeted tests for the purpose of regression testing of the software [9].

The major benefit of the technique is the risk analysis. After the change that is made in the software there may be potential risks. Test relevant to these risks should not be neglected. Those defects may result in catastrophic consequences. Furthermore, it provides high efficiency and it is cost effective as well. This technique is evaluated on a large industrial based case study, which is yet another plus point of the approach.

(2) Towards Software Architecture Based Regression Testing

In this approach [13] it is described that how regression testing can be applied systematically at software architecture level. This approach is used to reduce the cost of the retesting modified system and to check the regression testability of the evolved system. Software architecture based regression testing may be used in both cases of development and during maintenance. But according to [13] the focus is on maintenance aspect. This work builds upon the general framework for software architecture based conformance testing. The goal of the approach is to test whether the system implementation works correctly. This approach is tool supported by the LISA tool, the C2 framework, the Argus-I environment and the UNIX diff utility. They applied this technique on a case study for the validity of the approach. This technique is cost effective and tool supported. This approach seemed well evaluated and well validated.

3.3 Model Checking Based Approaches

In this section approaches are classified as model checking based approaches to generate regression test. One regression testing approach is found using model checking which is discussed below.

(1) Regression Testing Via Model Checking

Lihua, Dias, and Richardson [5] have proposed an approached for the regression testing via model checking. They named their approach as Regression Testing via Model Checking (RTMC). At the abstract level the basic idea of their approach is to take two specifications and find the differences between those specifications for the purpose of regression test generation. Both the original and the modified specifications are taken and passed to a comparator. The work of the comparator is to extract the differences between the original and modified specifications. From these differences the changed properties are identified called Extracted Prop. In the next step, all these properties are checked via a model checker to identify which properties of the specification are changed and which are not changed. Next, for the changed properties the test cases (for regression testing purpose) are generated and the specification is tested again. [5]

All the activities of this approach are performed through various components of the RTMC model. The components are RTMC controller, Comparator, Coverage Selector, Translator, and Test generator [5].

This approach supports different coverage criteria, includes a case study for the validation of technique. The technique is cost effective but not evaluated upon a large case study furthermore the efficiency of approach is not so significant.

3.4 Hybrid Approaches

The approaches which used multiple models for generating regression test cases are classified as hybrid approaches and discussed below.

The approach given by Farooq et al [4] is a model based selective technique for the regression testing based on the class diagram and state diagram model of UML.

For the purpose of the regression testing the changes in the modified version of the software is to be identified so that appropriate test cases should be generated to check whether the new version is fit for the purpose and the changes did not cause any problem for the unchanged portions of the software. To identify changes they [4] categorized the changes into two classes; Class driven and state driven changes.

As for as Class driven changes are concerned they identified these changes from three major categories of changing attributes i.e. operations, and relationship. the class driven changes they identified are ModifiedExpression, ChangedMultiplicity, ModifiedProperty, ChangedMultiplicity, ModifiedAttribute, ModifiedOperationParameter, ModifiredOperation, Modifiend Association, Added/deleted Attribute, Added/deleted Operation, Added/deleted association [4].

The second kind of the changes identified were state-driven changes. These changes were identified from the state diagram of the software design.

The state driven change categories identified were added/deleted state, modified state, added/deleted transition, modified transition, modified event, modified actions, and modified guards.

After the identification of these changes, test cases can be generated according to the categories of both classes of changes, which are in fact the test suite for regression testing.

They applied their technique on a case study for the validity of the approach. [4].This approach presents a case study to validate the technique. Their coverage criterion is high. It is yet to be evaluated on a large case study and needs not only to be automated but also to be made more efficient.

This is an approach [10] for the regression testing based on the UML models. To generate the regression test, three types of models from UML models are used for the purpose of the identification of the changes; class diagram, sequence diagram and use case diagram. On the basis of the identified changes from these UML models, the impact of the changes is also analysis to generate regression test suite [10].

First of all the change analysis is done in the two versions of class diagrams, i.e. original class diagram and the modified class diagram. There are different changes related to the class diagrams. The authors have identified that the changes may be added/deleted attribute, change attribute, added/ deleted method, changed method, added/deleted relationship, changed relationship added/deleted class and the changed class [10].

The next step is to identify changes in the two versions of sequence diagrams, original sequence diagram and the modified sequence diagram. Author has identified changes in two forms i.e. use case and the methods. Changes which may occur are add/deled use case, changed use case, add/deleted method and changed method [10].

On the basis of the impact of these changes, the test cases from the original suite are classified into three categories. 1) Obsolete test cases are those not required for the regression testing. 2) Retest able test cases are those which are required for the regression testing and will be included in regression testing. 3) Reusable test cases are those which are for the unaffected portion of the models.

The author has presented a regression test selection tool for this approach named as RTSTool [10].The technique is safe, efficient, cost effective and tool supported. The technique is also evaluated on a large industrial based case study.

Pilkasn et al [11] provided an approach for the regression testing based on the UML models as well. This approach is used to test the UML design model and to check the inconsistencies. A model is made which merges information from class diagrams, sequence diagram and OCL statements and further generates test cases to check inconsistencies. Furthermore, proposed approach categorizes changes found in UML design and then classifies the test cases which are based on categories of UML design.

They provide a set of rules i.e. how to reuse the existing test cases and how to generate the new test cases. New test cases will ensure that affected part of the system do not harmfully affect other parts of the software. The technique is safe, efficient, and cost effective. The case study is included but it is yet to be evaluated on a large case study.

4 Analysis of the Approaches

For evaluating different aspects of the model based regression testing approaches, we have identified some attributes including some attributes by Rothermel and Horrald [14]. A brief introduction of the parameters is presented and table 1 shows their acronyms, short definitions and possible values against any approach. In table 2 the acronym along with the classification of the approaches is given while in table 3 all the approaches discussed an analyzed in section 3, are evaluated upon the following parameters. Table 3 the core part of the research indicating the areas of research and improvements.

4.1 Safety

One very important question during the analysis of any regression testing approach is its safety that whether the technique is safe of not. This parameter is the part of the framework for evaluation of regression testing approaches provided by Ruthermel and Horrald [14].

4.2 Different Coverage Criteria

Some techniques are specific to only some coverage criteria, it is very important to be known for the improvement of the approaches. If an approach is not covering a specific criterion then the goal should be to extend the approach to make it possible that it should cover as many criteria as possible.

4.3 Efficiency

The landmark parameter in evaluating any kind of algorithm is its efficiency, so we are interested to set a value for the efficiency of each approach.

4.4 Risk analysis

It is more worthy if a technique is also based on the risk analysis of the system. It is necessary for the effectiveness and fault tolerant of the system.

Table 1 Evaluation Parameters

Acronym	Comparison attributes	Possible values of attributes
SF	Is technique safe?	Yes(Y), No(N)
DCC	Whether it satisfies different coverage criteria?	Yes(Y), No(N)
EFF	How much technique is efficient?	Low(below30) mod(between30to69) high(over70)
RB	Technique is risk based or not?	Yes(Y), No(N)
CE	Technique is cost effective or not?	Low(below30) mod(between30to69) high(over70)
CC	How much coverage criteria are satisfied?	Low(below30) mod(between30to69) high(over70)
CS	Whether case study is included or not?	Yes(Y), No(N)
AT	Technique is automated or not?	Yes(Y), No(N)
SC	Is technique evaluated on industrial case study?	Yes(Y), No(N)

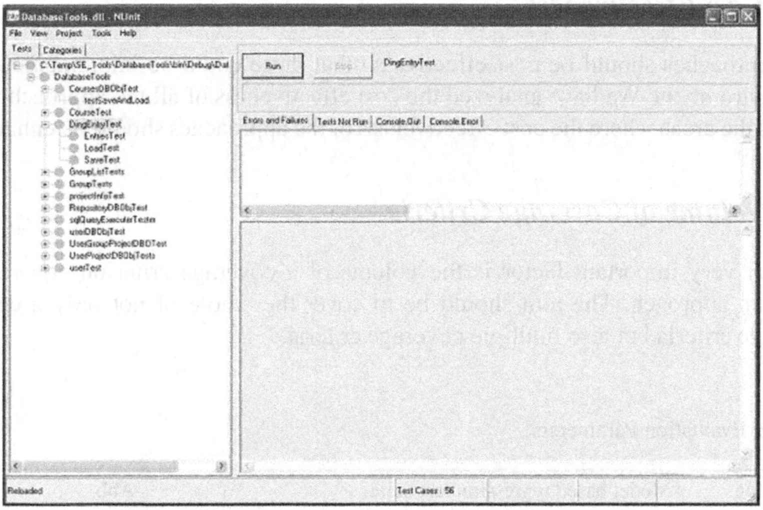

Fig. 1 Embedded Testing UI using MDA

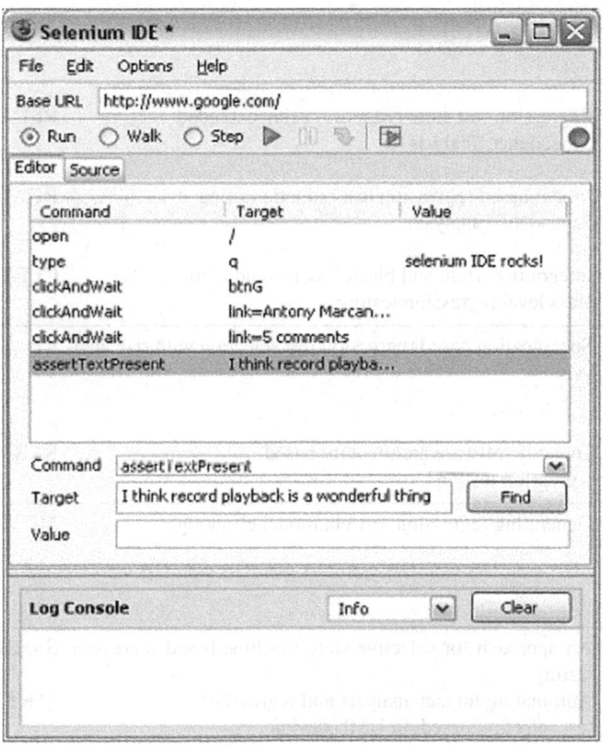

Fig. 2 Testing Results in this work

4.5 Cost Effectiveness

The approaches should be cost effective so that these could be implemented in a real environment. We have analyzed the cost effectiveness of all the approaches for finding the areas where the cost effectiveness of the approaches should be enhanced.

4.6 Volume of Coverage Criteria

Another very important factor is the volume of a coverage criterion covered by a certain approach. The aim should be to cover the whole of not only a certain coverage criteria but also multiple coverage criteria.

Table 2 Evaluation Parameters

Group	Model based regression test suite generation using dependence analysis	Abb.
State Machine Based Appoach	Model based regression test suite generation using dependence analysis	RTDA
	Regression test suite reduction using extended dependence analysis	REDA
	Model based regression test reduction using dependence analysis	RDA
	Integrating white and black box technique for class level regression testing	CLT
Activity Diagram Based Approach	Specification based regression test selection with risk analysis	RB
	Towards software architecture based regression testing	SAB
Model Checking Based Approach	Generating regression test via model checking	MC
Hybrid Approaches	An approach for selective state machine based regression testing	SSMB
	Automating impact analysis and regression test selection based on UML design	IART
	Regression testing UML design	UMLD

5 Examples

It is always worthwhile that the techniques should included a case study with them so that their worth in context of efficiency, coverage criteria, safety etc could be analyzed. Figure 1 shows what a typical test class would look like in this paper. Each method with the [Test] fixture would be its own test case and when run would have the [SetUp] run before it and the [TearDown] run after. The NUnit GUI is shown in figure 3; each leaf in the tree is a method with the [Test] fixture. If a tester wants they can run the whole test suite by clicking on the root, just a section of the test suite, or a single test case [4].

(1) Automation

Automation of the approaches depicts that it is validated to a larger extent and implemented. Furthermore the automation of the approaches leads it to be implemented in real world environment.

(2) Checking Scalability

One of most important parameters is checking the scalability of the approaches, none of the approaches could make its place in the industry without checking its scalability worth. The approaches must be evaluated on a larger industrial based case study before their implementation to the real world environment.

6 Conclusion

Automated testing is here to stay. Many project managers are faced with fast shrinking schedules and budgets to stay competitive in todays software market. Customers still demand top notch quality in the software that they buy though. This means that you need to keep the quality of your product high while doing it with less. Automated testing is expensive to implement, but when done right, can save huge amounts of time and money. In this paper, We presents the model based regression testing and their analysis with respect to the identified parameters.. It can be helpful in exploring new ideas in the area of regression testing specifically model based regression testing. In table 3 we have analyzed the techniques by some parameters we have identified. This evaluation of the model based regression testing techniques can be helpful to improve the existing techniques where they lack. This evaluation can also be very helpful to evaluate and upcoming technique. The parameters are not specific for only the model based regression testing but also can be helpful for evaluating and improving other regression testing approaches. One future recommendation is the improvements in the framework of Rothermel and Horrald [14] by including the parameters identified during this research.

References

1. Borland.: Borland SilkTest: An Automated Regression and Functional Software Testing Tool, `http://www.borland.com/us/products/silk/silktest/index.html` (retrieved April 2, 2007),
2. Dustin, E., Rashka, J., Paul, J.: Automated Software Testing: Introduction, Management, and Performance. Addison-Wesley, Reading (1999)
3. Lewis, E.W.: Software Testing and Continuous Quality Improvement. CRC Press LLC, Boca Raton (2005)
4. NUnit.: Nunit, `http://www.nunit.org` (retrieved April 2, 2007)
5. OpenQA.: OpenQA: Selenium, `http://www.openqa.org/selenium/index.html` (retrieved April 2, 2007)
6. Dalal, S.R., Jain, A., Karunanithi, N., Leaton, J.M., Lott, C.M., Bellcore, G.C.: Patton Model Based Testing in Pratice Software Engineering. In: Proceedings of the International Conference, pp. 285–294 (1999)
7. Nancy, Y.L., Wahi, J.: An overview of regression testing. ACM SIGSOFT Software Engineering Notes 24(1), 69–73 (1999)
8. Leung, H.K.N., White, L.: Insights into regression testing. In: Proc. IEEE International Conference on Software Maintenance (ICSM), pp. 60–69 (1989)
9. Farooq, Q., Iqbal, M.Z.Z., Mailk, Z.I., Nadeem, A.: An Approach for selective state machine based regression testing. In: Proceedings of the 3rd International Workshop on Advances in Model-Based Testing, London, United Kingdom, pp. 44–52 (2007)
10. Xu, L., Dias, M., Richardson, D.: Generating regression tests via model checking. In: Proceedings of the 28th Annual International Computer Software and Applications Conference (COMPSAC 2004), vol. 1, pp. 336–341 (2004)
11. Chen, Y., Probert, R.L., Ural, H.: Model-based regression test suite generation using dependence analysis. In: Proceedings of the 3rd International Workshop on Advances in Model-Based Testing, pp. 54–62 (2007)
12. Korel, B., Tahat, H.L., Vaysburg, B.: Model based regression test reduction using dependence analysis. In: Proceedings of the International Conference on Software Maintenance, ICSM 2002 (2002)
13. Chen, Y., Probert, R.L., Ural, H.: Regression test suite reduction using extended dependence analysis. In: Fourth International Workshop on Software Quality Assurance: in Conjunction with the 6th ESEC/FSE Joint Meeting, pp. 62–69 (2007)
14. Chen, Y.L., Robert, L., Probert, D., Sims, P.: Specification based regression test selection with risk analysis. In: IBM Centre for Advanced Studies Conference, Proceeding of the Conference of the Center for Advanced Studies on Collaboration Research (2002)
15. Briand, L.C., Labiche, Y., Soccar, G.: Automating impact analysis and regression test selection based on UML designs. In: Proceedings of the International Conference on software Maintenance (ICSM 2002). IEEE, Los Alamitos (2002)
16. Pilskan, O., Uyan, G., Andrews, A.: Regression testing UML design. In: Proceedings of the 22nd IEEE International Conference on Software Maintenance, pp. 254–264 (2006)
17. Beydeda, S., Gruhn, V.: Integrating white- and black- box techniques for class-level Regression Testing. In: Proceedings of the 25th International Computer Software and Applications Conference on Invigorating Software Development, pp. 357–362 (2001)
18. Muccini, H., Dias, M.S., Richerdson, D.J.: Towards software architecture-based regression testing. In: SESSION: Workshop on Architecting Dependable Systems (WADS), pp. 1–7 (2005)

An Automatic Architecture Reconstruction and Refactoring Framework

Frederik Schmidt, Stephen G. MacDonell, and Andrew M. Connor

Abstract. A variety of sources have noted that a substantial proportion of non trivial software systems fail due to unhindered architectural erosion. This design deterioration leads to low maintainability, poor testability and reduced development speed. The erosion of software systems is often caused by inadequate understanding, documentation and maintenance of the desired implementation architecture. If the desired architecture is lost or the deterioration is advanced, the reconstruction of the desired architecture and the realignment of this desired architecture with the physical architecture both require substantial manual analysis and implementation effort. This paper describes the initial development of a framework for automatic software architecture reconstruction and source code migration. This framework offers the potential to reconstruct the conceptual architecture of software systems and to automatically migrate the physical architecture of a software system toward a conceptual architecture model. The approach is implemented within a proof of concept prototype which is able to analyze java system and reconstruct a conceptual architecture for these systems as well as to refactor the system towards a conceptual architecture.

Keywords: Architecture reconstruction, software migration, source code transformation and refactoring, search based software engineering, metaheuristics.

1 Introduction

Software systems evolution is triggered by changes of business and technical requirements, market demands and conditions [1]. This evolution leads often to an erosion of the architecture and design of the system [2]. Reasons for this are several – insufficient time for design improvements [3], no immediate architecture

Frederik Schmidt · Stephen G. MacDonell · Andrew M. Connor
SERL, Auckland University of Technology
Private Bag 92006, Auckland 1142
New Zealand
+64 9 921 - 8953
e-mail: {fschmidt,stephen.macdonell,andrew.connor}@aut.ac.nz

R. Lee (Ed.): Software Eng. Research, Management & Appl. 2011, SCI 377, pp. 95–111.
springerlink.com

maintenance and management, fluctuation of employees, and different levels of knowledge and understanding of the conceptual architecture. Additionally, the automatic inclusion of imports within current IDE's obfuscates the creation of unwanted dependencies which conflict with the conceptual architecture. This erosion of design leads to fragile, immobile, viscous, opaque and rigid software systems [4]. Architecture reconstruction tools can help to create views of the physical architecture of the system [5], however these views do not fully support the ability to reconstruct a conceptual architecture view of the system which can be utilized as a blueprint for further development. Architecture management tools [6] such as *Lattix, Sotograph* and *SonarJ* help to monitor the design drift. To apply these, however, a clear conceptual architecture has to be defined. These tools offer only limited support to realign the conceptual and physical architectures [6]. Manual recreation of the conceptual architecture is hindered as the design erosion obfuscates the original intent of the system design. Additionally, existing architecture violations introduced during the system erosion have to be resolved to realign the physical architecture with the conceptual architecture. To avoid, or at least simplify, this complex and time consuming manual process we introduce an automatic framework for reconstruction and refactoring. This framework features the reconstruction of a conceptual architecture model based on acknowledged design principles and the resolution of architecture violations by applying software migration techniques. This facilitates the migration of the physical architecture model towards the conceptual architecture model. Before the framework is described, related work on software architecture reconstruction and migration is presented.

2 Related Work

The objective of this work relates to a variety of software engineering disciplines such as software architecture modeling, software architecture reconstruction, software architecture transformation and automatic refactoring of legacy systems. Strategies of software clustering and search based software engineering are applied to realise the objective of our study. Additionally, software architecture metrics are applied to evaluate the value of the generated solutions. To highlight the relevance and significance of our study relevant and contributing work of the named areas are illustrated in this section.

2.1 Software Architecture

The architecture of a software system visualizes a software system from an abstract view. An architectural view of a system raises the level of abstraction, hiding details of implementation, algorithms and data representations [7]. These architectural views can focus on different aspects of the system for example a service, implementation, data or process perspective [5]. Associated fine grained entities are classified into more abstract modules. Having a current representation of the system architecture is crucial in order to maintain, understand and evaluate a large software application [8].

Murphy & Notkin [9] depict the reflexion model as a framework to prevent software architecture deterioration of the implementation perspective. The reflexion model features a conceptual architecture which defines a desired model of the system and a mapping of the physical implementation units into the subsystems of the conceptual architecture model. The conceptual architecture model facilitates a development blueprint for the development stakeholders. An ideal conceptual architecture models the domain and technical environment of the software system and delivers a framework to maintain desired quality aspects.

The approach adopted by Murphy & Notkin [9] demands regular compliance checking of the physical and conceptual architectures and the immediate removal of architecture violations by applying refactoring techniques [10]. Therefore the documentation of a conceptual architecture and compliance checking of the conceptual and physical architecture is an important aid to maintain and understand software systems [11]. Within many software projects the compliance checking of physical architecture and conceptual architecture as well as the inclusion of realignment of the physical and conceptual architecture is not consequently included into the development process [12]. Additionally, software systems evolve due to requirement and framework changes during development. This may require altering the conceptual architecture. Consequently, the conceptual and physical architecture drifts apart without a rigorous compliance checking and refactoring of the physical architecture. The manual reconstruction of the architecture as well as the manual realignment is complex and time consuming.

2.2 *Software Architecture Reconstruction*

One of the challenges associated with architecture reconstruction is that often the available documentation is incomplete, outdated or missing completely. The only documentation of the software system is the source code of the system itself.

Reverse engineering activities help to obtain abstractions and views from a target system to help the development stakeholders to maintain, evolve and eventually re-engineer the architecture. The main objective of software architecture reconstruction is to abstract from the analyzed system details in order to obtain general views and diagrams with different metrics associated with them. Even if the source code might be eroded it is often the only and current documentation of the software system. Therefore the extraction of architecturally significant information and its analysis are the key goals which have to be determined to apply software architecture reconstruction successfully.

A variety of approaches and tools evolved to support the reconstruction of software architectures. Code Crawler [13] allows reverse engineering views of the physical architecture. The tool is based on the concept of polymeric views which are bi-dimensional visualisations of metric measurement such as *Lines of Code, Number Of Methods, Complexity, Encapsulation* etc. These views help to comprehend the software system to identify eroded and problematic artefacts. The Bauhaus [14] tool offers methods to analyze and recover the software architecture views of legacy systems; it supports the identification of re-usable components and the estimation of change impact.

In reverse engineering, software clustering is often applied to produce architectural views of applications by grouping together implementation units, functions, files etc. to subsystems that relate together. Software clustering refers to the decomposition of a software system into meaningful subsystems [15]. The clustering results help to understand the system. The basic assumption driving this kind of reconstruction is that software systems are organised into subsystems characterised by high internal cohesion and loose coupling between subsystems. Therefore, most software clustering approaches reward high cohesion within the extracted modules and low coupling between the modules [16]. Barrio [17] is used for cluster dependency analysis, by using the Girvan–Newman clustering algorithm to extract the modular structure of programs. The work of Macoridis & Mitchell [18, 19, 20] identifies distinct clusters of similar artefacts based on cohesion and coupling by applying a search based cluster strategy. These approaches are appropriate if the purpose is merely the aggregation of associated artefacts into a first abstraction of the system to redraw component boundaries in software, in order to improve the level of reuse and maintainability. Software architecture reconstruction approaches apply software clustering approaches to determine an architecture model of the system.

Chrstl & Koschke [21] depict the application of a search based cluster algorithm introduced in [22] to classify implementation units into a given conceptual architecture model. In a set of controlled experiments more than ninety percent of the implementation units are correctly classified into subsystems. The results indicate that an exclusively coupling based attraction function delivers better mapping results than the approach based on coupling and cohesion. Due to the given conceptual architecture the search space (clusters and dependencies between clusters) is distinctly reduced and a fair amount of the tiring process of assigning implementation units into subsystems is automated. However, it would be interesting if the approach is still feasibility if the erosion of the system is more pronounced. There is a high chance that with further erosion of the system that the error ratio would accumulate. Additionally, to apply the approach of Christl & Koschke [21] a conceptual architecture has to be evident to conduct the clustering. But as illustrated in the previous section and supported by various sources [7, 12, 23] current software systems face especially that the conceptual architecture is completely or at least partially lost.

The work of EAbreu et al. [16] complements the results of Christl & Koschke [21], showing that clustering based on the similarity metric and rewarding cohesion within subsystems and penalising coupling between subsystems does not provide satisfactory results which go beyond the visualisation of cohesive modules such as dependencies between modules, which would allow to model concepts as machine boundaries and encapsulation of modules.

The reconstruction of an architectural model, which can later be used as a conceptual architecture for further development is accompanied by two main problems, which cannot be solved with an approach which exclusively relies on maximises cohesion and minimising coupling based on a similarity function. The first problem is that a natural dependency flow from higher subsystems to modules of lower hierarchy levels exists. This dependency flow induces the cohesion and coupling based cluster algorithms to include artefacts of lower modules. Secondly, an architecture reconstruction is probably applied when the conceptual architecture is lost. Therefore a high degree of erosion might be evident in the physical architecture and correspondingly the assumption that a high internal cohesion and loosely coupling is evident might not be existent. Hence, to reconstruct an architectural model which fulfils the requirements of an architectural model more refined analysis techniques have to be applied. Other approaches base their analysis on non source code formations such as symbolic textual information available in comments, on class or method names, historical data (time of last modification, author) [24]. Other research includes design patterns as an architectural hint [25]. Sora et al. [8] enhance the basic cohesion and coupling based on the similarity and dissimilarity metric by introducing the concept of abstraction layers. Sora et al. [8] proposes an partitioning algorithm that orders implementation units into abstract layers determined by the direction of the dependencies. Sora et al. [8] do not include the possibility of unwanted dependencies. Therefore, architecture violating dependencies might bias the analysis and a higher degree of erosion leads to a solution with lower quality. Further evolved architecture reconstruction approaches aim to recover the layering of software systems as a more consistent documentation for development [26, 8].

Current approaches show the feasibility to reconstruct architectural views of software systems, however these approaches do not evaluate if these results are applicable to improve the understandability of the system or if the results are applicable as a conceptual architecture as part of the reflexion model. Additionally the illustrated architecture reconstruction approaches struggle to identify metrics beside cohesion and coupling to capture the quality of a conceptual architecture. The illustrated approaches do not consider that the physical architecture features a degree of erosion and deterioration which biases the reconstruction of a conceptual architecture. Current architecture reconstruction approaches create an abstract view of the physical architecture of the software system into. These abstraction views itself do not benefit a quality improvement of the system. They rather deliver a blue print for development stakeholders to understand the current implementation of the system. This enhanced understanding of the system can be utilised to conduct refactorings to improve the physical design of the system. Thus, the reconstruction of a conceptual architecture without changing the physical architecture will not improve the quality of the software system. Especially if the conceptual architecture has been reconstructed based on the source code of an eroded software system refactoring is required to realign the eroded design with the new conceptual architecture model. Hence, to improve the quality of the system the physical architecture has to be refactored in conjunction with the conceptual architecture to improve the overall design of the system.

2.3 Automatic Architecture Refactoring

One of our objectives of this study is to automatically realign the physical architecture with a given or reconstructed conceptual architecture. We understand the resolution of architecture violations as a migration of the physical architecture model to a new instance of the physical architecture model which features the same behaviour but aligns to a given conceptual architecture model. Therefore, work which feature automatic refactoring, architecture realignment and migration of software systems is of particular interest.

Refactoring is the process of changing the design of a software system by preserving the behavior [27]. This design improvement should positively benefit software quality attributes [10] such as testability, modularity, extendibility, exchangeability, robustness etc. Gimnich and Winter [28] depict migration as an exclusively technical transformation with a clear target definition. The legacy system is considered as featuring the required functionality and this is not changed by applying the migration. Therefore, the refactoring of a software system can be understood as a migration of a software system to another version which fulfils other quality criteria. Hasselbring, et al. [29] describe architecture migration as the adaptation of the system architecture e.g. the migration from a monolithic system towards a multi-tier architecture. Heckel et al.[30] illustrates a model driven approach to transform legacy systems into multi-tier or SOA architecture by applying the four steps *code annotation, reverse engineering, redesign and forward engineering*. The code annotation is the manual equipment with a foreseen association of architectural elements of the target system, e.g., GUI, application logic or data conducted by the development stakeholders [30]. The remaining three stages are executed guided by the annotations. If the identified solution is not satisfying the approach is iteratively repeated.

Ivkovic & Kontogiannis [1] propose an iterative framework for software architecture refactorings as a guideline to refactor the conceptual architecture model towards *Soft Quality Goals* using model transformations and quality improvement semantic annotations. The first step of Ivkovic & Kontogiannis [1] approach requires determining a *Soft Goal* hierarchy. The Soft Goal hierarchy is a set of *Soft Goals* ordered by relevance. The *Soft Goal* model assigns metric configurations to the *Soft Goals* high maintainability, high performance and high security. In the second phase a set of candidate architectural refactorings are selected which lead to improvements towards one of the *Soft Goals*. In the third stage the derived refactorings are annotated with compatible metrics which measure quality aspects of the concerned *Soft Goal*. Metric values are determined before and after conducting the refactoring to establish if the refactoring has a positive effect onto the quality attribute of the *Soft Goal*. Finally, the refactorings are iteratively conducted by selecting each soft goal of the soft goal hierarchy and implementing the refactorings with the highest benefit based on the previous metric measurements. O'Keeffe and Cinnéide [31] propose an automatic refactoring approach to optimize a set of quality metrics. They developed a set of seven complementary pairs of refactorings to change the structure of the software system. Metaheurisitc algorithms such as *multiple ascent hill-climbing, simulated annealing* and *genetic algorithm* are then used to apply the implemented refactorings. The fitness function to evaluate the refactored source code instance employs an implementation of the Bansiya's QMOOD hierarchical design quality model [32]. The QMOOD model comprises eleven weighted metrics depending on the weighting of these metrics the software quality attribute understandibility, reusability and flexibility can be expressed as a numerical measurement [32]. O'Keeffe and Cinnéide [31] utilizes these three different weightings as different fitness functions to refactor a system towards the desired quality attributes. They found that some of the example projects can be automatically refactored to improve quality as measured by the QMOOD evaluation functions. The variation of weights on the evaluation function has a significant effect on the refactoring process. The results show that first-ascent hill climbing produces significant quality improvements for the least computational expenditure, steepest-ascent hill climbing delivered the most consistent improvements and the simulated annealing implementation is able to produce the greatest quality improvements with some examples. O'Keeffe and Cinnéide [31] go on to state that the output code of the flexibility and understandability produced meaningful outputs in favour of the desired quality attributes where the reusability function was not found to be suitable to the requirements of search-based software maintenance because the optimal solution includes a large number of featureless classes.

3 An Architecture Reconstruction and Refactoring Framework

This section describes an automatic architecture reconstruction and transformation process designed to support the reconstruction of a conceptual architecture model of a software system and the migration of the analysed software system towards a given conceptual architecture model.

In the previous section a variety of architecture reconstruction, refactoring and migration approaches have been reviewed. It has been shown that current architecture reconstruction approaches are feasible to extract views of the physical architecture. The reconstructed architectural views can help development stakeholders to understand the current design of the system. However, the approaches are not aiming to reconstruct a conceptual architecture of the system or a blue print of the system which can be used for further development and compliance checking. Consequently, the re-creation of a conceptual architecture remains a tedious manual process which requires analyzing domain and technical environment aspects in compliance with the evident software system. One of the main problems while reconstructing a conceptual architecture is the erosion which might be evident in the system and bias the extraction of a conceptual architecture. The identification of violating dependencies is hard due to the uncertainty of the system deterioration. Automatic refactoring approaches refactor architectural views [1] or the source code [31] of the system towards predefined quality goals. These quality goals are represented as combinations of metrics which measure the desired quality goal. Migration and transformation approaches transform legacy systems into multi-tier or SoA architectures [29]. Most approaches require a substantial part of manual classification [30] hence a good understanding of the system is required. Other approaches transform views of the architecture without transforming the source code of the system [28]. Furthermore, approaches either transform or reconstruct architectural views or change the source code of the system. To our current understanding none of the reviewed approaches aim to provide a conceptual architecture view as well as a corresponding physical architecture model. We believe that a conceptual architecture model as well as a violation free physical model is one of the key requirements to enable an active and continuous architecture management. Additionally, the evidence of a corresponding reflexion model delivers the base for further development and refactoring of the system towards better architectural design.

Based on this we propose and evaluate a combination of architecture reconstruction techniques to extract a conceptual architecture model and refactoring techniques to obtain an aligned physical architecture and conceptual architecture model. The process reflects that a conceptual architecture model can be based on acknowledged software design principles represented as architecture styles, design patterns and software metrics. The conceptual architecture model represents a target definition for the migration of the physical architecture model. The reconstruction and transformation process is outlined in Figure 1 which illustrates the input and output relationships.

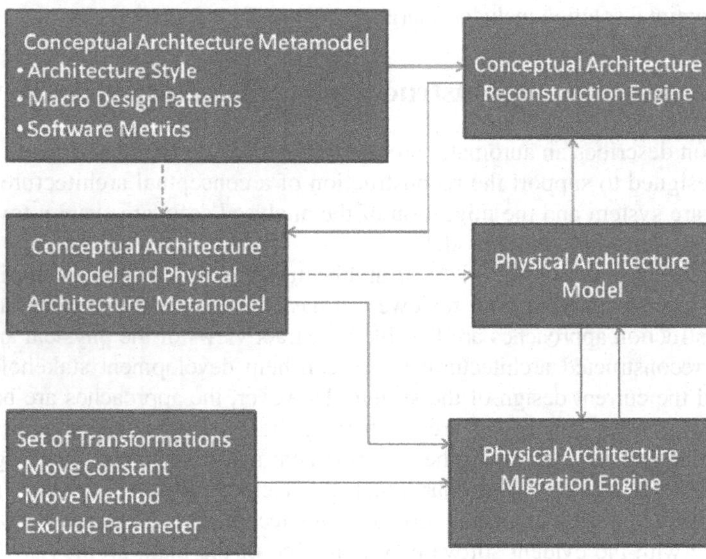

Fig. 1 Automatic Architecture and Migration Framework

A prototype has been developed to enable an evaluation of the feasibility of the framework. This prototype allows the reconstruction of a conceptual architecture as well as the refactoring of the physical architecture towards this conceptual architecture model on the basis of java software systems. Dependency analysis as well as the migration of source code instances are enabled by applying the RECODER source code optimisation and transformation framework [33]. The following two sections illustrate architecture reconstruction and architecture migration as implemented in our framework.

3.1 Architecture Reconstruction

An automatic conceptual architecture reconstruction framework is useful if the desired architecture of a system is lost. As previously stated, the manual reconstruction of conceptual architecture in fairly eroded systems is complex and labour intensive [34]. The objective of this component is to evaluate if the automatic reconstruction of a reflection model is feasible in terms of delivering a modular structure of a software system as a basis for further development. The reconstructed architecture model delivers a conceptual blueprint of the system that implements established design principles of software architectures. This blueprint can be used in further development, refactoring and architecture monitoring.

Considering the necessity to apply an architecture reconstruction it can be assumed that no or only limited documentation of the conceptual architecture exists and the physical architecture of the system is eroded. Hence, regular compliance checking has not been conducted due to the missing basis for comparison. Additionally, it is also not possible to determine the degree of erosion as the basis for comparison in the form of a defined design description is missing and the erosion is influenced by a variety of factors such as development activity, design drift, design understanding of development stakeholders, framework changes and time without compliance checking. Additionally, the requirements of an ideal conceptual architecture candidate change over time caused by requirement, environment and framework changes. The definition of an ideal conceptual architecture depends on variables such as the development environment, development philosophy, applied frameworks and functional requirements. It is hard to capture all these variables within an automated approach based on the analyses of legacy code. However, we are convinced that at least having a conceptual architecture has long term benefits on the life cycle and quality of the software system.

We suggest a search based cluster approach to reconstruct a conceptual architecture. This decision is based on the complexity of the problem, the size of the search space and also the multiplicity of optimal solutions. To date the reconstruction of a layered architecture style with transparent horizontal layers has been implemented. A search based clustering algorithm similar to the clustering approach of Mitchell and Mancoridis [20] classifies the implementation units into n layers. As an acknowledged software clustering concept the clustering penalizes high coupling between clusters and rewards high cohesion within the clusters.

We employ a greedy metaheurisitic to identify a start solution and apply a steepest ascent hill climbing metaheuristic to improve this initial solution. Our approach utilizes the work of Harman [35] which states that metrics can act as fitness functions. Our objective is to recreate a system architecture that exhibits good modularity.

We designed a fitness function *Solution Quality* to evaluate the fitness of a solution. Based on the *Soft Goal* graph of Ivkovic & Kontogiannis [1] we utilize the *Coupling Between Objects* metric as measurement for modularity. Every dependency between implementation units is annotated with a *CBO* measurement. Our greedy algorithm classifies the implementation units into clusters based on rewarding cohesion and penalizing coupling. The clusters are ordered ascending based on the ratio of incoming and outgoing dependencies. Additionally, we reward solutions with more layers. The *Solution Quality* is multiplied with the number of layers in the system. However, at this stage we only allow solutions with three or less layers. The steepest ascent hill climbing algorithm tries to increase the *Solution Quality* measurement by swapping implementation units between clusters.

In our model an architecture violation is a dependency from a lower layer to a higher layer. These dependencies deteriorate the encapsulation of two layers if we take the conceptual architecture as an optimal solution. As the system probably features an indefinite degree of deterioration we do not just want to minimize the number of architecture violations. Hence, just relying on the minimization of architecture violations would model the deteriorated system into a conceptual model and therefore not challenge an improvement of the system design. Our overall aim is to obtain a violation free architecture of the system. To support this approach we classify between violations which can be resolved by the automatic refactoring (defined in sections 3.2.1 and 3.2.2) and violations our approach is not capable to resolve. Each dependency is tested if it can be resolved by one of our three automatic refactoring transformations. If a dependency can be resolved by the application of refactoring the CBO weight of the dependency is multiplied with a factor of 0.25 and therefore the dependency is rather an accepted architecture violation as it does only increase the coupling between layers by a small degree. Hence, we penalize the inclusion of architecture violations which cannot be resolved with our by multiplying the CBO measurement with a factor of 2.0, which strongly increases the coupling between layers and therefore penalizes the solution.

The output of this conceptual architecture reconstruction is a conceptual architecture model which comprises ordered layers and implementation units which are mapped into these layers. Therefore, a reflexion model has been created. However, the physical architecture might feature dependencies which violate with the reconstructed conceptual architecture model [9].

3.2 Automatic Architecture Refactoring

This section describes an automatic refactoring framework to migrate the physical architecture towards a given conceptual architecture by applying a set of transformations.

Our automatic refactoring approach expects as input the reflexion model of a software system. This reflexion model can be user-defined or can be created by our previously illustrated architecture reconstruction method. The objective of this component is to deliver an automatic source code transformation approach that has the potential to re-establish the modularity of eroded software.

The migration framework aims to resolve architecture violations which cannot be resolved by reclassifying implementation units into different subsystems of the conceptual architecture model. The origin of these architecture violations is hidden in the implementation of the system, which does not align with the conceptual modularity and decomposition of the system [7]. To resolve architecture violations of this kind the source code of the implementation unit has to be migrated to comply with the conceptual architecture. A set of allowed transformations had to be specified to migrate the system from one instance to another.

The objective of the automatic refactoring is the resolution of unwanted dependencies between implementation units. This requires the application of refactorings which alter the dependency structure between implementation units. Rosik, Le Gear, Buckley, Babar and Connolly [2] found that a significant number of violations are based on misplaced functionality. Another common reason for the erosion of design is the injection of higher-classified implementation units as a parameter and access to the structure and behaviour of these objects from lower-classified implementation units [12]. Therefore the three transformations *move method*, *move constant* and *exclude parameter* have been implemented within our proof-of-concept prototype. These transformations refactor the implementation unit which causes the architecture violation as well as the interrelated callers of the refactored code element. The behaviour of these transformations is as follows:

3.2.1 Move Method and Move Constant

These transformations move a method or constant that causes an architecture violation to any other implementation unit. As it cannot be assumed to which implementation unit the refactored code artefact should be moved, the code artefact in question is placed into every implementation unit of the system. For each of these outcomes a new instance of the source code is created as the base for the application of further transformations.

3.2.2 Exclude Parameter

The exclude parameter transformation excludes one parameter of a method. The code elements which reference or access the parameter are moved to the caller implementation units. Currently the order of execution can be changed by applying this transformation and consequently the program behaviour might change. Our current solution is to exclude the parameter and instead include a listener pattern which notifies the top layer implementation unit to execute the refactored code elements. Based on an identified architecture violation one of the three

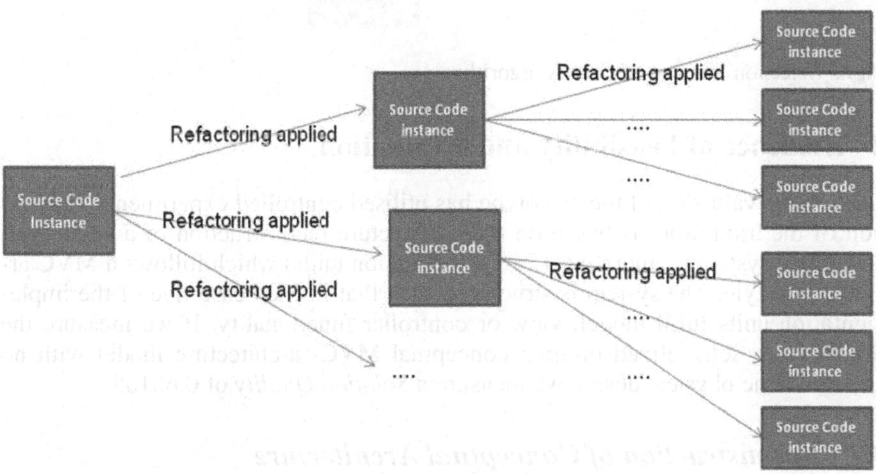

Fig. 2 Evolution of Source code instances based on applied transformations

implemented transformations can be selected. A new instance of the software system is created based on the application of every transformation. The complexity of an exhaustive search would quickly result in an uncontrollable number of generated software system instances. Figure 2 illustrates the uncontrolled generation of software system instances.

Due to this computational complexity we apply a greedy algorithm to control the reproduction process of software system instances. Based on the initial solution a population of new software instances is created. The fittest solution is selected based on the lowest number of architecture violations as the primary selection criteria. If two solutions feature the same number of architecture violations the selection is based on the fitness function illustrated in section 3.1. Figure 3 illustrates the selection strategy and reproduction for two generations.

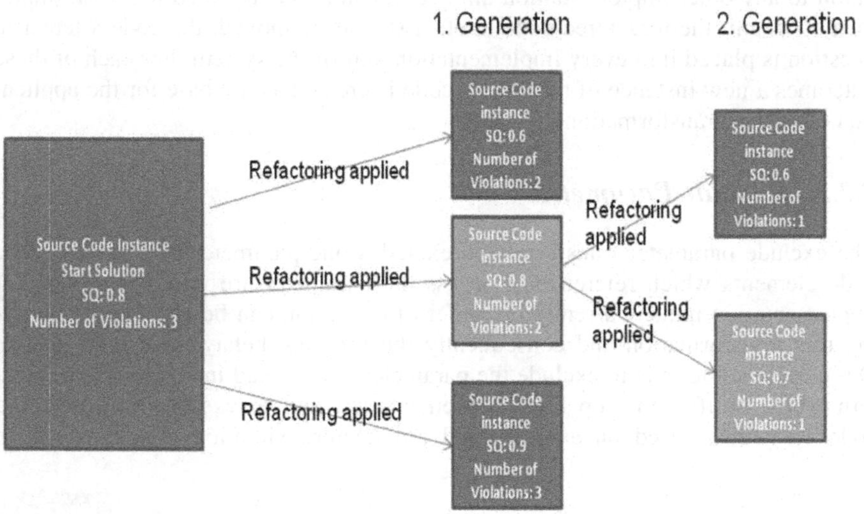

Fig. 3 Selection Strategy of Greedy algorithm

4 Evidence of Feasibility and Evaluation

Our initial evaluation of the prototype has utilised controlled experiments. Evaluation of the framework is based on the architecture reconstruction of a small self-developed system (comprising 15 implementation units) which follows a MVC architecture style. The system is structured such that in each case five of the implementation units fulfil model, view or controller functionality. If we measure the fitness for a self defined optimal conceptual MVC- architecture model with no erosion in the physical design we measure a *Solution Quality* of 0.66185.

4.1 Reconstruction of Conceptual Architecture

A set of 11 experiments has been conducted, in each experiment an architecture violating dependency is added and then resolution is attempted. These imposed architecture violations conflict with the MVC architecture. The type of violation is equally distributed by adding wrongly placed methods, constants and wrongly injected parameters. The results of the experiments are shown in Table 1.

Table 1 Results Architecture Reconstruction Experiment

No. Injected Architecture Violations	No. Layers	Misplaced implementation units	Architecture violations	Solution Quality
0	3	2	0	0.64835
1	3	2	0	0.63464
2	3	2	1	0.53571
3	3	3	2	0.55833
4	3	3	3	0.45889
5	3	3	3	0.42652
6	2	5	2	0.58534
7	2	5	2	0.58534
8	2	6	3	0.54532
9	2	6	5	0.53345
10	2	6	6	0.52345

The results show that the prototype is able to identify a conceptual architecture model of the system and classify the implementation units into corresponding layers. The *Solution Quality* as a fitness representation tends towards lower values with increasing erosion of the system. The only break in this general trend is the reduction of layers in the conceptual architecture to two which causes disjointed values.

In each of the experiments the resulting conceptual architecture features a set of implementation units which are not classified correctly and also the number of identified violations differs from the number of initiated architecture violations; hence the identified architecture violations are not necessarily identical to the introduced architecture violations. The experiments also show that, based on the rising number of misplaced implementation units, the constructed conceptual architecture model drifts further from the initial MVC architecture. However, at this stage the suggestion of a conceptual architecture model which can be used as a base for further development and refactoring to regain a degree of system modularity seems to be feasible.

4.2 Realignment of Physical Architecture with Conceptual Architecture Model

We conducted a second set of experiments based on the physical architecture of our self developed MVC example. We utilised the physical design with 10 injected architecture violations and the reconstructed conceptual architecture model with 2 layers from our previous experiments. The objective of these experiments is to evaluate the automatic refactoring of the physical architecture towards a given conceptual architecture model.

To evaluate the feasibility of the automatic refactoring towards a given conceptual architecture it is necessary to evaluate if the process contributes to a quality improvement of the software system. The proposed approach aims to re-establish the modularity of the system. The main objective of the automatic refactoring is the reduction of architecture violations. However, the number of architecture violations depends strongly on the given conceptual architecture model. So far the *Solution Quality* fitness function is available to evaluate the modularity of the system. Table 2 shows the results of this experiment.

Table 2 Results Architecture Refactoring Experiment

Generation	Number of Architecture Violations	Solution Quality
1	6	0.52345
2	5	0.50432
3	4	0.54545
4	3	0.56362
5	2	0.53756
6	2	0.53756

A reduction of architecture violations can be observed during the first five generations. From the fifth generation no appropriate move can be identified to resolve the remaining architecture violation. The *Solution Quality* fitness function measurement reflects no significant quality improvement of the refactored system and no clear trend of the *Solution Quality* measurement can be recognised. The reason for this might be the individual evaluation of architecture violations in the model in respect to their resolvability with our implemented refactorings. To evaluate the quality of the generated solution more general metrics should be applied to allow estimating the overall quality development of the system.

In general, it has been found that violations based on wrongly placed constants can be completely resolved. The outcome of resolutions using the move method and exclude parameter transformations depends on the dependencies of the method and parameter to the initial containing implementation unit. If no interrelation to the containing implementation unit exists the method can be placed into other implementation units or the parameter can be excluded and the violation resolved. However, these preliminary results show that a migration from one instance of a software system to another is feasible by applying a set of defined transformations which align the software system with a given conceptual architecture model.

5 Limitations

The conducted evaluation is preliminary but is encouraging. Further application in real scenarios is necessary (and is ongoing) to more fully assess the applicability of our Architecture Reconstruction and Migration Framework.

6 Conclusions and Future Work

This paper describes a framework designed to reconstruct a conceptual architecture model for legacy systems and to migrate the physical architecture model of legacy systems towards a given conceptual architecture model. Based on the theoretical illustration of the causes and consequences of deteriorated software design, the possibility to utilise the conceptual architecture model as a metamodel for the physical architecture is illustrated. The method of operation of the architecture reconstruction by utilizing acknowledged macro-architecture design principles and the physical architecture model is described. Furthermore the principles of operation of the software migration framework by utilising the conceptual architecture model, applying design patterns, software metrics and source code transformation are described. Finally, preliminary results of our feasibility evaluation are presented and discussed.

At this time our prototype addresses a limited set of architecture styles and transformations. We are working to extend the number of possible architecture styles by introducing vertical layering and impervious layers to model functional decomposition and machine boundaries in the conceptual architecture. Further research will also focus to resolve architecture violations by the migration of the source model towards design patterns. Another current working area is the extension of the *move method* and *exclude parameter* transformations to migrate interrelations to containing implementation units. The current search strategies are immature. It will be beneficial especially for the evaluation of larger software systems to guide the search towards more promising solution candidates by applying other search strategies e.g *genetic algorithms*. Future work will involve evaluating if the migration approach has the potential to migrate a software system from one conceptual architecture model to another. This is of particular interest if the conceptual architecture of a system changes due to requirement, environment and technology changes and the conceptual architecture model and mapping of implementation units into this new architecture can be defined.

References

[1] Ivkovic, I., Kontogiannis, K.: A framework for software architecture refactoring using model transformations and semantic annotations. In: Proceedings of the 10th European Conference on Software Maintenance and Reengineering (CSMR 2006), Bari, Italy, p. 10 (2006)

[2] Rosik, J., Le Gear, A., Buckley, J., Babar, M.A., Connolly, D.: Assessing architectural drift in commercial software development: A case study. Software: Practice and Experience 41(1), 63–86 (2010)

[3] Kerievsky, J.: Refactoring to patterns. Addison-Wesley Professional, Reading (2005)

[4] Martin, R.C.: Design principles and design patterns (2000), Object Mentor retrieved from `http://www.objectmentor.com/resources/articles/Principles_and_Patterns.pdf`

[5] Koschke, R.: Architecture reconstruction: Tutorial on reverse engineering to the Architectural Level. In: International Summer School on Software Engineering, pp. 140–173 (2008)

[6] Telea, A., Voinea, L., Sassenburg, H.: Visual tools for software architecture understanding: A stakeholder perspective. IEEE Software 27(6), 46–53 (2010)
[7] Bass, L., Clements, P., Kazman, R.: Software architecture in practice. Addison-Wesley Professional, Reading (2003)
[8] Sora, I., Glodean, G., Gligor, M.: Software architecture reconstruction: An approach based on combining graph clustering and partitioning. In: Proceedings of the International Joint Conference on Computational Cybernetics and Technical Informatics, pp. 259–264 (2010)
[9] Murphy, G.C., Notkin, D.: Reengineering with reflexion models: A case study. IEEE Computer 30(8), 29–36 (1997)
[10] Fowler, M.: Refactoring: Improving the design of existing code. Addison-Wesley Professional, Reading (1999)
[11] Ducasse, S., Pollet, D.: Software architecture reconstruction: A process-oriented taxonomy. IEEE Transactions on Software Engineering 35(4), 573–591 (2009)
[12] Van Gurp, J., Bosch, J.: Design erosion: problems and causes. Journal of Systems and Software 61(2), 105–119 (2002)
[13] Lanza, M., Ducasse, S., Gall, H., Pinzger, M.: Codecrawler: an information visualization tool for program comprehension. In: Proceedings of the 27th International Conference on Software Engineering, pp. 672–673 (2005)
[14] Raza, A., Vogel, G., Plödereder, E.: Bauhaus – A Tool Suite for Program Analysis and Reverse Engineering. In: Pinho, L.M., González Harbour, M. (eds.) Ada-Europe 2006. LNCS, vol. 4006, pp. 71–82. Springer, Heidelberg (2006)
[15] Wiggerts, T.A.: Using clustering algorithms in legacy systems remodularization. In: Proceedings of the Fourth Working Conference on Reverse Engineering (WCRE 1997), Amsterdam, Netherlands, October 6-8, vol. 43, pp. 33–43 (1997)
[16] Abreu FB, Goulão M, Coupling and cohesion as modularization drivers: Are we being over-persuaded?. Proceedings of the 5th Conference on Software Maintenance and Reengineering, Lisbon, Portugal (2001)
[17] Dietrich, J., Yakovlev, V., McCartin, C., Jenson, G., Duchrow, M.: Cluster analysis of Java dependency graphs. In: Proceedings of the 4th ACM Symposium on Software Visualization, Herrsching am Ammersee, Germany, September 16-17, pp. 91–94 (2008)
[18] Mancoridis, S., Mitchell, B.S., Chen, Y., Gansner, E.R.: Bunch: A clustering tool for the recovery and maintenance of software system structures. In: Proceedings of the IEEE International Conference on Software Maintenance (ICSM 1999), Oxford, England, UK, August 30-September 3, pp. 50–59 (1999)
[19] Mitchell, B.S., Mancoridis, S.: On the automatic modularization of software systems using the Bunch tool. IEEE Transactions on Software Engineering 32(3), 193–208 (2006)
[20] Mitchell, B.S., Mancoridis, S.: On the evaluation of the Bunch search-based software modularization algorithm. Soft Computing-A Fusion of Foundations, Methodologies and Applications 12(1), 77–93 (2008)
[21] Christl, A., Koschke, R., Storey, M.A.: Equipping the reflexion method with automated clustering, 10–98 (2005)
[22] Mitchell, B.S.: A heuristic search approach to solving the software clustering problem. PhD, Drexel University, Drexel (2002)
[23] Fontana, F.A., Zanoni, M.: A tool for design pattern detection and software architecture reconstruction. Information Sciences: An International Journal 181(7), 1306–1324 (2011)

[24] Andritsos, P., Tzerpos, V.: Information-theoretic software clustering. IEEE Transactions on Software Engineering 31(2), 150–165 (2005)
[25] Bauer, M., Trifu, M.: Architecture-aware adaptive clustering of OO systems. In: Proceedings of the Eighth Euromicro Working Conference on Software Maintenance and Reengineering (CSMR 2004), pp. 3–14 (2004)
[26] Scanniello, G., D'Amico, A., D'Amico, C., D'Amico, T.: An approach for architectural layer recovery. In: Symposium on Applied Computing 2010 (SAC 2010), Sierre, Switzerland, pp. 2198–2202 (2010)
[27] Opdyke, W.F.: Refactoring: A program restructuring aid in designing object-oriented application frameworks. PhD thesis, University of Illinois at Urbana-Champaign (1992)
[28] Gimnich, R., Winter, A.: Workflows der Software-Migration. Softwaretechnik-Trends 25(2), 22–24 (2005)
[29] Hasselbring, W., Reussner, R., Jaekel, H., Schlegelmilch, J., Teschke, T., Krieghoff, S.: The dublo architecture pattern for smooth migration of business information systems: An experience report. In: Proceedings of the 26th International Conference on Software Engineering (ICSE 2004), pp. 117–126 (2004)
[30] Heckel, R., Correia, R., Matos, C., El-Ramly, M., Koutsoukos, G., Andrade, L.: Architectural Transformations: From Legacy to Three-Tier and Services. In: Software Evolution, pp. 139–170. Springer, Heidelberg (2008), doi:10.1007/978-3-540-76440-3_7
[31] O'Keeffe, M., Cinnéide, M.Ó.: Search-based software maintenance. In: Proceedings of the 10th European Conference on Software Maintenance and Reengineering (CSMR 2006), pp. 249–260 (2006)
[32] Bansiya, J., Davis, C.G.: A hierarchical model for object-oriented design quality assessment. IEEE Transactions on Software Engineering 28(1), 4–17 (2002)
[33] Ludwi, A.: Recoder homepage (March 22, 2010), `http://recoder.sourceforge.net`
[34] O'Brien, L., Stoermer, C., Verhoef, C.: Software architecture reconstruction: Practice needs and current approaches. Technical Report CMU/SEI-2002-TR-024, Carnegie Mellon University (2002)
[35] Harman, M., Clark, J.: Metrics are fitness functions too. In: Proceedings of the International Software Metrics Symposium, Chicago, Illinois, USA, September 14-16, pp. 58–69 (2004)

Modelling for Smart Agents for U-Learning Systems

Haeng-Kon Kim

Abstract. Mobile software development challenges the modeling activities that precede the technical design of a software system. The context of a mobile system includes a broad spectrum of technical, physical, social and organizational aspects. Some of these aspects need to be built into the software. Selecting the aspects that are needed is becoming increasingly more complex with mobile systems than we have previously seen with more traditional information systems. Mobile computing poses significant new challenges due the disparity of the environments in which it may be deployed and the difficulties in realizing effective software solutions within the computational constraints of the average mobile device. In this paper, we discuss the creation of such a model and its relevance for technical design of a smart agent for u-learning mobile software system. Conventional approaches to modeling of context focus either on the application domain or the problem domain. These approaches are presented and their relevance for technical design and modeling of software for agent mobile systems is discussed. The paper also reports from an empirical study where a methodology that combines both of these approaches was introduced and employed for modeling of the domain-dependent aspects that were relevant for the design of a software component for mobile agents. We also discuss some pertinent issues concerning the deployment of intelligent agents on mobile devices for certain interaction paradigms are discussed and illustrated in the context of a u-learning applications.

Keywords: Smart Agent, U-learning, Mobile Systems, Component Based Systems, Embedded System, Software Modeling, Model Driven Architecture.

1 Introduction

Recent progress of wireless technologies has initiated a new trend in learning environments called ubiquitous learning (u-learning) which is able to sense the situation

Haeng-Kon Kim
Department of Computer Engineering, Catholic University of Daegu, Korea
e-mail: hangkon@cu.ac.kr

R. Lee (Ed.): Software Eng. Research, Management & Appl. 2011, SCI 377, pp. 113–128.
springerlink.com

of the learners and automatically offer corresponding supports even can provide personalized services to anything, by anyone, at anytime and anywhere. Nowadays, U-learning environment has been various designed and implemented. The concept of ubiquitous learning (u-learning) has become a real possibility over recent years as those technologies inherent to the ubiquitous computing paradigm mature. A practical realisation of this may be seen with ever increasing penetration of mobile devices into society and the increased availability and capacity of public telecommunication networks. In effort to make information more relevant, particularly in the WWW sphere, significant research has taken place in the area of personalisation and the use of intelligent collaborative agents. By combining these technologies with an open-source course management system, a personalised u-learning system for the delivery of initial entrepreneurial training has been created. During initial requirements gathering, it became clear that new entrepreneurs are time-limited so a ubiquitous learning experience would be most beneficial for such students. Also, entrepreneurs tend to have diverse educational backgrounds and the issue of linguistic diversity was also of importance. Ubiquitous computing presents challenges across computer science: in systems design and engineering, in systems modeling, and in user interface design.

Education has undergone major changes in recent years, with the development of digital information transfer, storage and communication methods having a significant effect. This development has allowed for access to global communications and the number of resources available to todays students at all levels of schooling. After the initial impact of computers and their applications in education, the introduction of e-learning and m-learning contains the constant transformations that were occurring in education. Now, the assimilation of ubiquitous computing in education marks another great step forward, with Ubiquitous Learning (u-learning) emerging through the concept of ubiquitous computing. It is reported to be both pervasive and persistent, allowing students to access education flexibly, calmly and seamlessly. U-learning has the potential to revolutionary education and remove many of the physical constraints of traditional learning. Furthermore, the integration of adaptive learning with ubiquitous computing and u-learning may offer great innovation in the delivery of education, allowing for personalisation and customisation to student needs[1].

In this study we discuss the creation of such a model and its relevance for technical design of a smart agent for u-learning mobile software system. Conventional approaches to modeling of context focus either on the application domain or the problem domain. These approaches are presented and their relevance for technical design and modeling of software for agent mobile systems is discussed. The paper also reports from an empirical study where a methodology that combines both of these approaches was introduced and employed for modeling of the domain-dependent aspects that were relevant for the design of a software component for mobile agents. We also discuss some pertinent issues concerning the deployment of intelligent agents on mobile devices for certain interaction paradigms are discussed and illustrated in the context of a u-learning applications.

2 Related Research

The majority of Web-enhanced courses reside on Learning Management Systems (LMS) [2]. LMSs provide on-line learning platforms that enable the consolidation of mixed media content, automate the administration of courses, hosts and deliver subject matter, integrate well with other applications, measures students learning effectiveness and manage the delivery and tracking of all electronic learning resources. Additional elements of an LMS include entry to a collaborative learning community, access to various formats of course materials, for example WWW, videos, audio, scheduling of on-line classes as well as group learning (forums and on-line chats). Examples of some of todays most utilized LMS include commercial systems like WebCT [3] and Blackboard [4] with other open source offerings such as Moodle.

A recent concept introduced to the education arena is that of ubiquity or U-learning. U-learning removes the confines of the physical classroom and creates an environment where remote education is available any time, any place, any media (Figure 1). One way in which a user profile can be maintained is through the use of smart agents. Agents have been used for personalisation in the form of context modelling and also to anticipate the needs of users. They provide a useful construct for which to automate the adaptation of content presentation to the requirements of an individual user, as well as the device and network they are using. Agent technologies are advancing at an exponential rate at present and their uses have been identified as fundamental to the development of next generation computing[5]. The utilisation of agents provides a powerful mechanism for automating services such as personalisation and content adaptation.

A model is an abstracted view of a system that expresses related knowledge and information. It is defined by a set of rules, used to interpret the meaning of its components[6]. A model is defined according to a modeling language that can give a formal or a semi formal meaning description of a system depending on modelers intention. The model paradigm has gained in importance in the field of systems engineering since the nineties. Its breakthrough was favoured by working groups like the Object Management Group (OMG)[7] that has normalized modeling languages such as Unified Modeling Language (UML) . This group also provides the Model-driven architecture (MDA) software design standard. The MDA is the main initiative for Model Driven Engineering (MDE) approach. According to OMG, four abstraction levels have to be considered: a meta-meta-model that represents the modeling language and which is able to describe itself; a meta-model level that represents an abstraction of a domain and which is defined through the meta-meta-model; a model level that gives an abstraction of a system as an instance of the meta-model; finally, the last abstraction level is the concrete system. In order to breath life in models [8], model transformations aims to exceed the contemplative model view to a more productive approach, towards code generation, analysis and test, simulation, etc. Models are transformed into other models.

Yet they have very little in particular to offer in modelling context for mobile systems. Rational Unified Process[, for example, offers several principles of which none address how to model the context of a mobile system. Microsoft Solutions

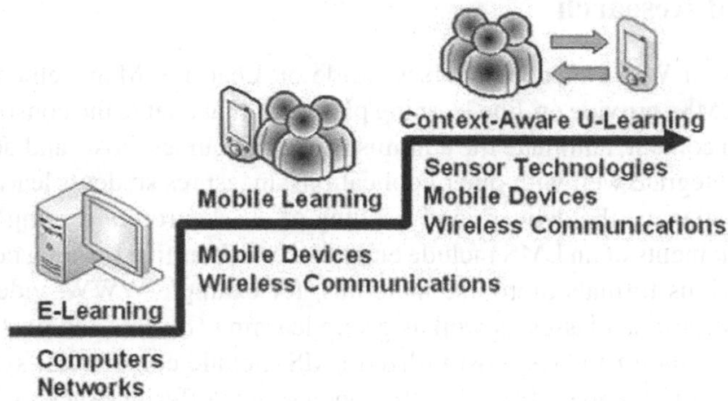

Fig. 1 The Ubiquitous Learning Computing Model

Framework[9], as another example, offers a set of principles for software engineering, but, again, has nothing in particular to say on modeling the context of a mobile system. The literature on human-computer interaction has a stronger emphasis on the context of computerized systems. The basic literature deals with user interface design from a general point of view, e.g.[10]. They provide extensive guidelines and techniques for user interaction design but nothing specific on design of mobile systems and very little on modelling of domain-dependent aspect as a basis for technical design. Some of the literature in human-computer interaction deals specifically with user interaction design for mobile systems. There is a general textbook on design of user interaction for mobile systems[11]. This textbook has a strong focus on mobile systems but significantly less on the modelling of domain-dependent aspects and very little emphasis on the relation to software engineering and technical design of software. There is also a frowing body of literature on context awarteness. Some of this literature discusses definitions of context and the implications for systems that are aware of their context. This group includes both conceptual work practical guidelines for modelling and design,[12]as well as examples of the process of analysing mobile activity and designing context-aware mobile systems. A common characteristic of this literature is that they work with various domain models, but there is very little about the relation to other models that are made for design of technical aspects, including the representation of information about the relevant domains. There is some literature that deals with technical design of context-aware systems. For example, Anagnostopoulos et al. formulate a set of requirements for what should be modelled in designing mobile systems: context data acquisition (e.g., from sensors), context data aggregation, context data consistency, context data discovery, context data query, context data adaptation, context data reasoning, context data quality indicators, and context data integration. Henricksen et al[13] start by recognizing that what is unique to modelling context for mobile systems: information on context is temporal, it is imperfect, context has several representations, and context information is highly interrelated. Thedy then go on to suggest a graphical modelling language that allow them to use three entity types people, device, and communication channel. In addition, they can model association between entities like static or temporal, dependencies between associations, and finally qualities of dependencies like accuracy and certainty. Lei and Zhang[14] apply the theory of conceptual graphs to mobile context modelling. All items in the context are monitored and the conceptual graphs model this as simple graphs associated with rules and constraints.

3 Modeling a Smart Multi Agent for U-Learning

The system provide a individualized service to learners in ubiquitous environment with smart multi agent for u-learning(SMAU). It also recognizes and changes the external environment automatically and the process contents with fuzzy techniques that will help to improve the results. As these users will have diverse needs and variable amounts of time to undertake the learning, we conceived a system that adjusts to their requirements and can be undertaken when and where suits the user. The SMAU system is constructed from a combination of the Moodle learning management system and a community of collaborative intelligent agents realized via Agent Factory as in figure 2.

3.1 Capturing Interaction through Agents

On completing requirements analysis and, as part of the initial design stage, the software designer, possibly in conjunction with a HCI engineer, must decide on the necessity for supporting the implicit interaction modality for the proposed application. Should they decide favorably, then a mechanism for monitoring and capturing both explicit and implicit interaction must be identified. A number of options exist; and the final decision will be influenced by a number of factors that are application domain dependent. However, one viable strategy concerns intelligent agents. From the previous discussion on interaction and intelligent agents, it can be seen that certain characteristics of the agent paradigm are particularly suited for capturing interaction. Practically all applications must support explicit interaction. The reactive nature of agents ensures that they can handle this common scenario. As to whether it is prudent to use the computational overhead of intelligent agents just to capture explicit user input is debatable, particularly in a mobile computing scenario. Of more interest is a situation where implicit interaction needs to be captured and interpreted.

Implicit interaction calls for continuous observation of the end-user. As agents are autonomous, this does not present any particular difficulty. Though identifying some particular aspect or combinations of the user's context may be quite straightforward technically, interpreting what constitutes an implicit interaction, and the appropriateness of explicitly responding to it in a timely manner may be quite difficult. Hence, the need for a deliberative component. As an illustration of the issues involved, interaction modalities supported by SMAU, a system based on the agent paradigm are now considered. SAMU is a functioning prototype mobile computing

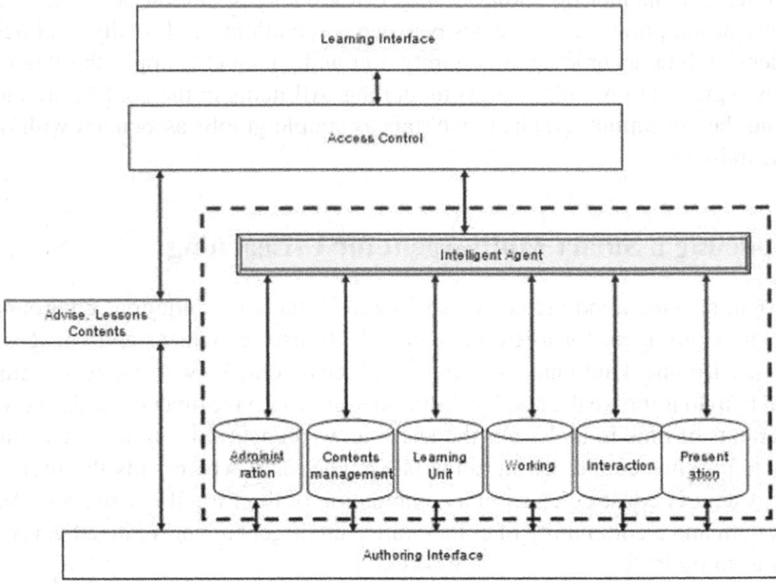

Fig. 2 SMAU System Architectures

application, developed to illustrate the validity of the u-learning paradigm. By augmenting u-learning with smart and autonomous components, the significant benefits of convenience and added value may be realized for the average learners as they wander their local street.

3.2 The SMAU Multi-Agent System

These agents are Administration Agent, Contents Management Agent, Learning Unit Agent, Working Agent, Interaction Agent and the Presentation Agent as in figure 3.

The Administration. This Agent detects user interactions with the SMAUe interface and records the results of these interactions. When the user first logs in, this Agent extracts the given username and password and passes this information to the Profiler Agent, which then accesses the users profile. When the user completes a questionnaire, quiz or other self-assessment exercise, this Agent passes the responses and results to the Profiler Agent. This information is used to build a profile of the users abilities and how they view their own experience to date.

The Profiler Agent. The Profiler Agent is responsible for maintaining, updating and analysing the user profile. This profile is then used to generate personalised content based on the users abilities, network capacity, device type and native language. The Profiler Agent uses a set of rules to assess the users requirements, and

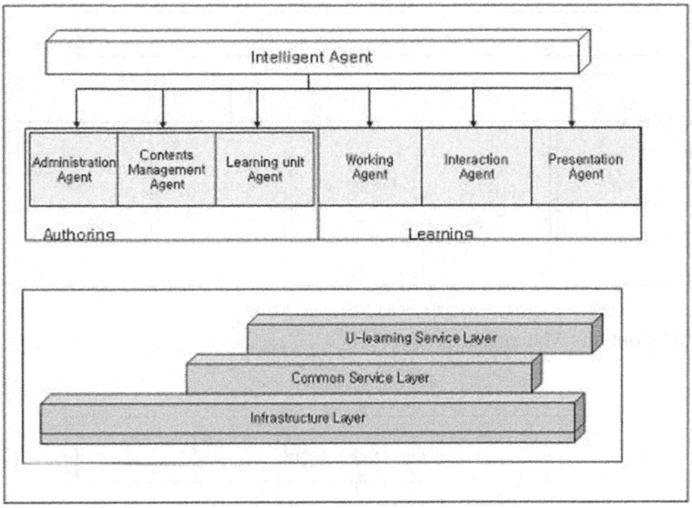

Fig. 3 SMAU Multi Agent Layers

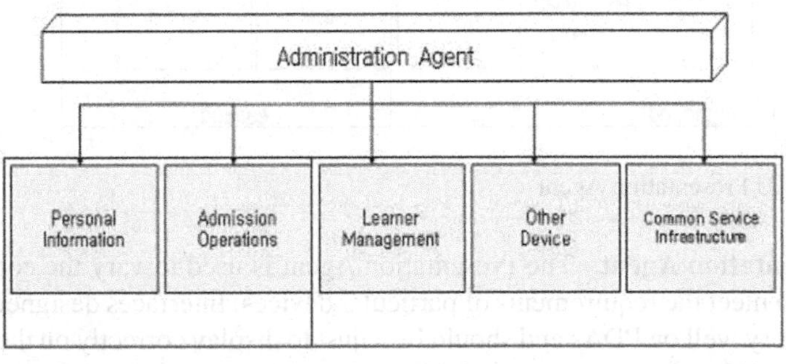

Fig. 4 SMAU Administration Agent

transmits these requirements to the Content Management and Presentation Agents. These rules govern the behaviour of the entire system and ensure that any personalisation is intuitive and does not appear intrusive to the user.

The Content Management Agent. The Content Management Agent controls the content to be passed to the user. When the Profiler Agent detects a particular content requirement, this information is passed to the Content Management Agent, which then decides which materials should be displayed. For example, if a user has performed particularly badly on an assignment the Profiler Agent will notify the Content Management Agent to re-display the preceding courseware so that the user may revise this information until they achieve the required competency to progress to the next level.

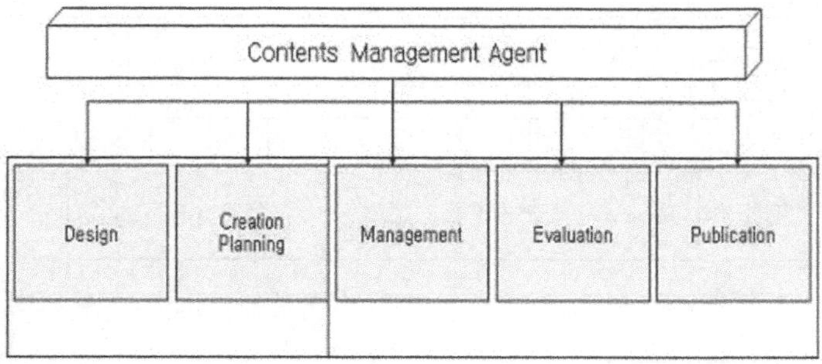

Fig. 5 SMAU Contents Management Agent

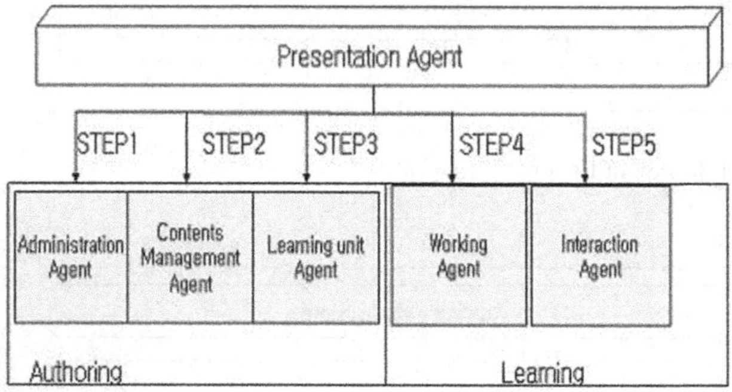

Fig. 6 SMAU Presentation Agent

The Presentation Agent. The Presentation Agent is used to vary the content and interface to meet the requirements of particular devices. Interfaces designed for PCs do not fit very well on PDAs and should be adjust to display correctly on the reduced screen size. This is important in the field of u-learning as users should have the same learning experience irrespective of where and when they wish to learn. By adjusting the content so that only the most relevant information is displayed, and adjusting or possibly removing the peripheral interface components, such as calendars etc., the user can still achieve their learning goals equally well on a PDA and PC.

The user creates their profile including preferred language during initial registration within Moodle. The Listener Agent acquires the given username and password and passes these to the Profiler Agent. The Profiler Agent can then use the username and password to poll the Moodle database for profile data, which it then stores within the User Profile Database. The Profiler Agent also implicitly appends acquired information to this explicitly provided profile to create a truly representative user profile. As the user navigates their way through Personalisation is a key factor in the creation of u-learning materials as the accepted model of u-learning is particularly user-centric. Personalised learning dictates that instruction should not be restricted by time, place or any other barriers and should be tailored to the continuously modified user model [10]. Entre-pass uses agents to automatically adjust the user profiles to meet these requirements as they are automatically detected based on the users interactions with the interface. The targeted users for Entre-pass are extremely varied in their educational backgrounds and learning needs, with the added challenge of linguistic diversity as the course-ware is currently available in 5 languages: English, Spanish, Danish, Romanian and Hungarian. In order to achieve adequate personalisation of the u-learning materials, a constantly adapting user profile is maintained and adjusted by the Profiler Agent the courseware, a record of their passage through this information space is maintained by the Listener Agent and this information is used by the Profiler Agent to constantly update the stored user profile.

This interaction record includes the results of all assessments such as quizzes and questionnaires and uses these to indicate whether the user has achieved the required capabilities to progress to the next level of the course. If not, the Profiler Agent will notify the Content Management Agent, which will re-issue the previously viewed material to the user so that they can revise that section of the course. Another vital piece of information that is implicitly detected is the IP and MAC address of the users device. This information is used to adapt the interface to a suitable format for the device in use. For example, the screen real estate varies widely between PDAs and desktop PCs so the cascading style sheet of the SAMU will adapt to a PDA format thus giving the user a more comfortable learning experience. Figure 6 show the overall architecture for SMAU. It is strongly related between learner and administration and agents.

4 Experience from the Modelling Process

This section presents the second part of results of our action case study. Here, the emphasis is on the experience that was gained when using the method to create the domain-depdendent models that were presented above.

The System Definition. The purpose of the system definition is to delineate the context of the system considered on an overall level. The following challenges were experienced with the system definition:

- Vehicle for keeping focus
- Defining granularity
- Point of view
- Application domain versus functions

It was difficult to make a good system definition for the smart mobile learning agents. The reason is that it expresses key decisions that are difficult to make. However, once it was made, it turned out to be a very powerful tool for keeping focus during the whole modelling activity. Sometimes, the participants forgot to use the system definition when they discussed an issue related to the limits of the problem or application domains.

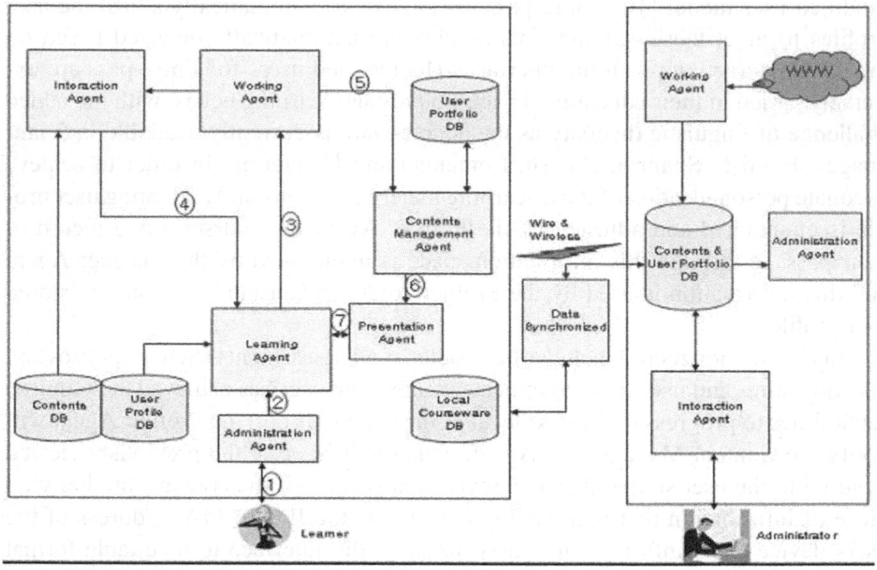

Fig. 7 Overall Architectures for SMAU

The system definition inherently involves defining the granularity in the model of the systems context. For example, the system definition mentions key objects, and this influences the considerations made during problem domain analysis. The participants had difficulties selecting a relevant level of granularity. They could either model a single component of the system or the whole system. When they focussed on the whole system, they also had a choice of seeing it as one or a few complex components or a larger number of simpler components. The solution employed was to see the system as one unit with relations to other systems.

SAMU system can always be described from two opposite points of view. Either from the outside where it is considered as a black box; or from the inside where the focus is on its components. For example, the participants could describe the relation between the user and the mobile smart agents, or they could describe leraner agent as a component and its interplay with other components. The participants shifted between these points of view. For the system definition, it was suggested to see the system as a black box and, therefore, not deal with its components.

The system definition has the application domain as one of its six elements. The aim of this is to emphasize who the users are and what their role is in the user organization. The participants had problems describing this element. In most cases there was no clear distinction between application domain and functions. This is also reflected in the system definition shown in figure 7. A mobile system is usually operated by a single user. Therefore, the application domain becomes very simple, and the developers had difficulties believing that it could be that simple.

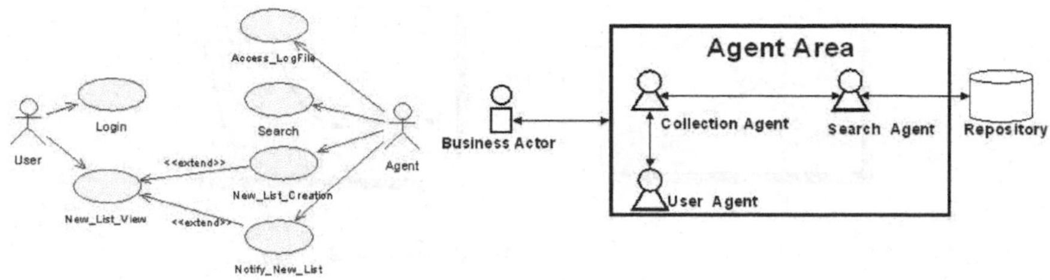

Fig. 8 System Sefinition for SAMU

This challenge emphasizes a key difference between conventional and mobile systems. For a stationary application, it is often useful to describe the domain where it is applied. For a mobile system, that is not possible because of the mobile nature of the system. This follows from the definition of a mobile system that was provided in the introduction. For a mobile system, it is more interesting to describe the application situations, i.e. the situations in which the device is used.

The Problem Domain Model. The problem domain model describes the part of the context that the users administer, monitor and control by means of the system. The modelling of this faced the following challenges:

- Identifying classes
- Identifying events
- No physical counterparts
- Classes with a single object
- Classes versus functions
- Users and clients
- Structural relations

Identification of the key classes was relatively straightforward. It was difficult to get started, but once the first candidates were identified, the participants suggested several more. The criterion used was whether the system would need to maintain information about objects from a class that was suggested. In addition, a class must be manageable. Some classes may be so small that they are really only representing an attribute. An example is Current Station. It is still included in the model because it represents a key concept that should appear in the model. We identified classes of a category that is not seen in conventional systems. The SAMU depends on two other components that are modelled as classes: It turned out to be much more difficult to identify events. We ended up modelling events that represent user actions and state changes for related components that the system needs to register or respond to. The identification of events involved significant discussion of the way in which the system can detect that these events have happened.

The model of the problem domain can employ the following object-oriented structures: generalization, aggregation and association. In our study, it was quite easy to explain these means of expression in general, but it was hard to relate them

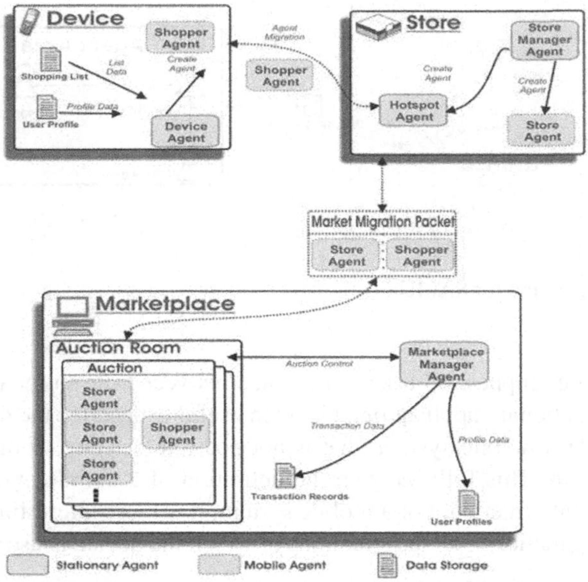

Fig. 9 SAMU Execution Environments

to their actual counterparts when the system was running. It was especially difficult with aggregation and association. This was solved by focusing on the has-a definition of aggregation. In addition, we used construction and destruction of objects to differentiate between associated and aggregated objects.

The Application Domain Model. The purpose of the application domain model is to describe how the model of the problem domain is applied by users and other systems. Key elements in the model are use cases and a complete list of system functions. The modeling of this faced the following challenges:

- Identifying users
- Generating use cases

The conventional approach is to define the users of the system as the people who are working in the application domain. Above, it was emphasized that it was impossible to identify a specific application domain for the SAMU. As a consequence, we had no simple way of identifying the users of the system. In addition, many components in a mobile system are applied by other components. Some of these components may then have users, and a difficult question was whether we should describe the component as the user or the user of that external component as the users of the system in question. To solve this, we introduced the notion of client which includes users as well as other systems or components. Thereby, we emphasized the direct user of the system considered.

A related problem was that it was difficult to generate use cases. They are usually defined by identifying tasks in the application domain and then selecting those that

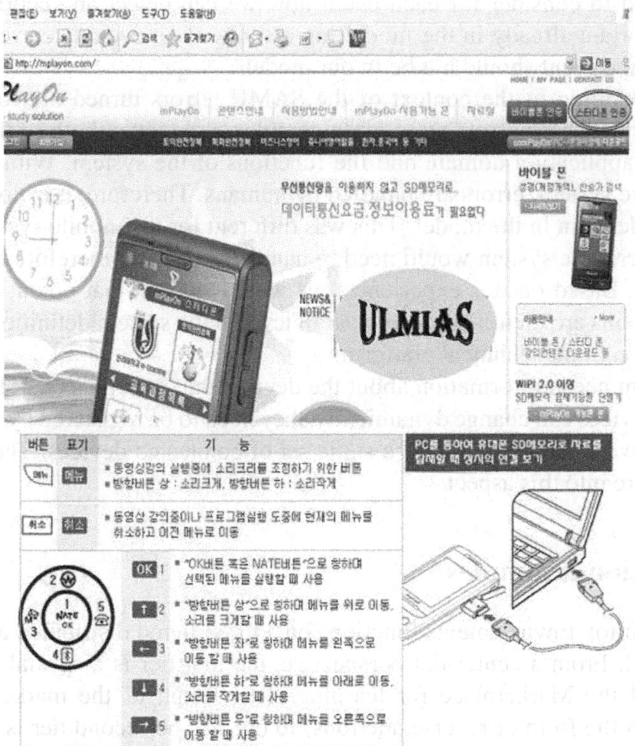

Fig. 10 SAMU Install Screen

will be supported by the system. This is basically a top-down approach. The difficulty of describing the application domain for the SAMU implied that we could not use that approach. Instead, we tried a bottom-up approach, where the participants described a set of activities that would involve use of the system.

For the system definition, it was suggested to focus on application situations instead of the application domain. This might also make it easier to generate relevant use cases. However, we did not try this idea out in our study.

Modelling System Behaviour. The problem domain model is by nature static. During the modelling of the context for the SAMU, certain dynamic issues were considered. These issues were:

- Detecting event occurrences
- Handling errors
- Connected devices

Events are described as part of the problem domain model. In design, it is decided how they are represented in the model and presented on the user interface. When the system is running, the model should be updated each time an event occurs. Therefore, we need to consider for each event how we can detect an occurrence. This issue is important already in the modelling activity. If an event occurrence cannot be detected, the event should not be in our model.

In the modelling of the context of the SAMU, errors turned out to be a very important issue. These errors were often complex and related both to the problem domain, the application domain and the functions of the system. With a classical administrative system, errors are handled by humans. Therefore, it is not as important to include them in the model. This was different for the mobile system that we modelled. Here, the system would need to handle errors, and therefore they should be described. Based on this experience, we would suggest that potential conflicts between systems are identified on an overall level in the system definition as part of the description of the technical platform.

The system needs information about the devices that are connected. If the set of connected devices can change dynamically, they need to be registered by the system. In our case, we turned out to have a static set of connected devices. Therefore, we did not inquire into this aspect.

5 Execution Examples

SAMU execution Environments functions on a three-tiered distributed architecture as in figure 8. From a center-out perspective, the first tier is a special centralized server called the Marketplace for learning. The design of the marketplace permits trade (in the form of reverse auctions) to occur. The second tier is termed the Hotspot. This is a hardware and software hybrid suite, situated at each participating retail outlet, which is permanently connected to the Marketplace and which allows the process of moving (migrating) representative (selling) agents from the retailers together with (buying) agents representing shoppers to the Marketplace. The final and outermost tier in this schema is collection of device nodes.

6 Conclusion

This paper addresses how we can model domain-dependent aspects of mobile systems with the purpose of providing a basis for designing the system. We have surveyed existing research literature on this question. The existing literature focuses either on the problem domain or the application domain. For designing a mobile system, this is not adequate. We have presented results from an empirical study of domain modeling in relation to a specific smat mobile agent software system. We have modeled the application and problem domains for this system with the aim of providing a useful basis for technical design of the mobile system. The application has been used in an action case study where we have enquired into teaching and deployment of a modelling method in a large group of software developers in mobile software companies. From the action case study we have presented both the results in terms of the models we arrived at and in terms of the process we went through. The resulting models show that the problem domain for the example case has been modelled in terms of classes and events. The process experience emphasizes the challenges faced by software developers whehn modeling domain-dependent aspects of a mobile system. The method we have applied in the action case study and reported on in this paper suggests modelling of domain-depdendent apsepcts of mobile systems based on ther following five principles:

- The modelling of context is based on a fundamental distinction between the problem domain and the application domain.
- The starting point is a system definition that assists in defining the systems scope and delimits the problem and application domains.
- The problem domain is modelled in terms of the basic concepts object and events and described in a class diagram and a set of statechart diagrams.
- The application domain is modelled in terms of use cases and functions and described in related specifications.
- The model and its selection of objects and events are useful for both design of representation and presentation in a software system.

References

1. RDoherty, B.C., O'Hare, P.T., O'Grady, M.J., O'Hare, G.M.P.: Entre-pass: Personalising u-learning with Intelligent Agents. In: Proceedings of the Fourth International Workshop on Wireless, Mobile and Ubiquitous Technologies in Education (WMUTE 2006), pp. 58–62. IEEE Computer Society, Athens (2006)
2. Brusilovsky, P.G.: A Distributed Architecture for Adaptive and Intelligent Learning Management Systems. In: AIED Workshop. Towards Intelligent Learning Management Systems (July 2003)
3. WebCT: Course Management System, Linfield, MA, WebCT INC (2002), `http://www.webct.com`
4. Blackboard, INC. Blackboard Course Management System (2002), `http://www.blackboard.com`
5. Luck, M., McBurney, P., Priest, C.: A Manifesto for Agent Technology: Towards Next Generation Computing. In: Autonomous Agents and Multi-Agent Systems, vol. 9, pp. 203–252. Kluwer Academic Publishers, Dordrecht (2004)
6. Gérard, S., Terrier, F., Tanguy, Y.: Using the Model Paradigm for Real-Time Systems Development: ACCORD/UML. In: Bruel, J.-M., Bellahsène, Z. (eds.) OOIS 2002. LNCS, vol. 2426, pp. 260–269. Springer, Heidelberg (2002)
7. Cranefield, S., Purvis, M.: UML as an ontology modelling language. In: Proceedings of the Workshop on Intelligent Information Integration, 16th International Joint Conference on Artificial Intelligence (IJCAI 1999), vol. 212 (1999)
8. Czarnecki, K., Helsen, S.: Classification of model transformation approaches. In: Proceedings of the 2nd OOPSLA Workshop on Generative Techniques in the Context of the Model Driven Architecture (2003)
9. Turner, M.S.V.: Microsoft Solutions Framework Essentials. Microsoft Corporation (2006)
10. Dix, A., et al.: Human-Computer Interaction. 3rd edn. Prentice-Hall, London (1998); Preece, J., Rogers, Y., Sharp, H.: Interaction Design: Beyond Human-Computer Interaction. John Wiley and Sons, New York (2002)

11. Morse, D., Armstrong, S., Dey, A.K.: The What, Who, Where, When, Why and How of Context-Awareness. In: CHI 2000. ACM, New York (2000)
12. Kjeldskov, J., Paay, J.: Augmenting the City: The Design of a Context-Aware Mobile Web Site. In: DUX 2005. ACM, San Francisco (2005); Anagnostopoulos, C.B., Tsounis, A., Hadjiefthymiades, S.: Context Awareness in Mobile Computing Environments. Wireless Personal Communications Journal (2006)
13. Henricksen, K., Indulska, J., Rakotonirainy, A.: Modeling Context Information in Pervasive Computing Systems. In: Pervasive 2002. Springer, Heidelberg (2002)
14. Lei, S., Zhang, K.: Mobile Context Modelling using Conceptual Graphs. In: Wireless And Mobile Computing, Networking and Communications (WiMob 2005). IEEE, Los Alamitos (2005)

Examining Software Maintenance Processes in Small Organizations: Findings from a Case Study

Raza Hasan, Suranjan Chakraborty, and Josh Dehlinger

Abstract. Software maintenance constitutes a critical function that enables organizations to continually leverage their information technology (IT) capabilities. Despite the growing importance of small organizations, a majority of the software maintenance guidelines are inherently geared toward large organizations. Literature review and case-based empirical studies show that in small organizations software maintenance processes are carried out without following a systemic process. Rather, they rely on ad-hoc and heuristics methods by organizations and individuals. This paper investigates software maintenance practices in a small information systems organization to come up with the nature and categories of heuristics used that successfully guided the software maintenance process. Specifically, this paper documents a set of best practices that small organizations can adopt to facilitate their software maintenance processes in the absence of maintenance-specific guidelines based on preliminary empirical investigation.

Keywords: Small organizations, software maintenance, case study, cognitive heuristics, ad-hoc process.

1 Introduction

Software maintenance is extremely important for organizations of all sizes; 60–80 percent of organizational resources are spent on maintenance as opposed to 20-40 percent on software development (Takang and Grubb 1996). Software maintenance is defined in IEEE Standard 1219 as "the modification of a software product after delivery to correct faults, to improve performance or other attributes, or to adapt the product to a modified environment" (IEEE 1993). There are four types of software maintenance: (Lientz and Swanson 1980): 1) Adaptive: Changes made to adapt to software environment; 2) Perfective: Changes made based on user

Raza Hasan · Suranjan Chakraborty · Josh Dehlinger
Towson University, USA
e-mail: {rhasan,schakraborty,jdehlinger}@towson.edu

R. Lee (Ed.): Software Eng. Research, Management & Appl. 2011, SCI 377, pp. 129–143.
springerlink.com

requirements; 3) Corrective: Changes made to fix errors/bugs; 4) Preventive: Changes made to prevent problems in the future. A vast majority of these changes (75%) are either adaptive or perfective. Corrective changes (fixing defects) comprise 20% of software maintenance of activities and preventive maintenance typically constitutes less than 5% of the total mix (Pigoski 1997).

Small organizations play a vital role on the world stage. Economic growth of many countries including U.S., Brazil, China and European countries rely heavily on small businesses (Software Industry Statistics 1991-2005). According to Process Maturity Profile, organizations that have 25 or fewer employees allocated to software development are considered small (Software CMM 2005).A critical challenge for small organizations in successfully implementing software maintenance functions is the lack of resources (Pigoski 1997). This results in insufficient processes, tools and documentation. Additionally, there is lack of adaptable process improvement models that can be specifically adopted by a small organization (April et al. 2005). Existing process improvement models (e.g. CMMI) are typically extremely resource intensive and are problematic for a small organization to adopt because of their prohibitive resource costs and lack of ease of implementation (Staples et al. 2007). Therefore most small organizations are inadequately prepared to meet the demands of comprehensive process frameworks. Despite the challenges faced by small organizations in carrying out software maintenance functions, there has been very little research investigating how such organizations perform their software maintenance operations (Brodman and Johnson 1994). Therefore, there is an imperative need to investigate the nature of software maintenance operations in small organizations. This article reports the preliminary finding of an exploratory investigation into the nature of software maintenance operations in small organizations using a qualitative case study based approach.

The rest of the paper is organized as follows. We first provide details of our case study based approach. Then we report our preliminary finding. Finally we conclude with a discussion of our findings.

2 The Case Study

We adopted an interpretive case study approach (Walsham 1995) to conduct the empirical study for this research. The choice of the methodological approach was predicated primarily by the context of our particular research focus. In this study, we were interested in obtaining an in depth understanding of a phenomenon (software maintenance processes within a small organization) that has not been explicitly investigated in information systems research. In addition our intention was to obtain a rich description and develop preliminary theoretical insights about the phenomenon in question. Case study research has been recommended as an appropriate methodology in situations where the "intention is to investigate a contemporary phenomenon within its real life context" (Yin, 1994, pg 13), the research and theoretical development understanding of the particular phenomenon are at a formative stage (Benbasat et al. 1987) and the "focus is on the in-depth understanding of a phenomenon and its context" (Darke et al. 1998). The site for the study was an Information Systems department of a university. This site has a total staff of sixteen people including seven developers, five analysts, three managers and one director. Its main focus is to meet the software development and software maintenance requirements of the university. It is a small organization, but its user base is quite large with over 20,000 students and more than 3,000 employees. A significant portion of work undertaken at this site comprises software maintenance. To get a good mix of data, three maintenance projects of varying sizes (i.e., large, medium and small) were chosen. Large projects represent those that are carried over a time period of more than twelve weeks, medium are those that are two to twelve weeks in duration and small projects require less than two weeks time. Table 1 below shows the details of these projects.

Table 1 Projects under Study

	Large	**Medium**	**Small**
Project Name	Portal	Financial Aid	Faculty Roles
Time Span	More than 12 Weeks	11 Weeks	1 Week
Software Maintenance Type	Adaptive and Perfective	Adaptive and Perfective	Perfective

The Portal project was implemented over a span of fourteen months and entailed enhancing the existing faculty and staff web portal with four major enhancements: Remove disjointed web pages, improve navigation, allow single sign-on, and provide personalization. This was a major project that involved adapting to user requirements and new technologies. The Financial Aid project was initiated to adjust a student's financial aid information per new legislation. This project was undertaken to fulfill legal requirements and provide new functionalities to users. The Faculty Roles Update project entailed making changes to existing program per new requirements of Registrar's and Human Resource Offices. The objective of this project was to make faculty access to student information more convenient.

Qualitative data for these three projects was collected from two groups: the staff of the information systems (IS) organization and end-users of applications. IS staff included two developers, two analysts and one manager. They were interviewed in-depth regarding their understanding of the term software maintenance, software maintenance processes, resources and tools, and enablers and inhibitors of software maintenance. These interviews were semi-structured and ranged from 40 – 60 minutes. In addition, nine interviews were conducted with end-users (faculty and staff) regarding their understanding of the enhancements needed in the Portal Project. Interviews with development team members were all tape recorded and transcribed. Interviews with the users were all written down only due to their reluctance to interview recording. Detailed notes of these user interview notes and transcribed notes of IS team were analyzed using an interpretive approach following the recommendations of leading qualitative scholars (Walsham 2006).

3 Preliminary Findings

Our initial analysis of the interviews suggests that the organization under study did not use any of the existing software maintenance process models. Instead we learned that the preferred approach was a Quick-Fix Model. This approach is essentially ad-hoc and does not offer specific guidelines; problems are fixed as they appear (Grubb and Takang 2003). The main disadvantage of this model is it does not consider ripple effects. This approach is not resource intensive and perhaps the easiest to implement, however it is not holistic and relies heavily on human actors and heuristics. Such reliance on heuristics and human actors brings along its set of complexities which makes it very challenging to implement software maintenance projects. As shown in Figure 1, software maintenance challenges faced by a small organization in software maintenance are lack of resources - personnel, time, and funds (Pigoski 1997). Lack of resources in-turn results in ad-hoc processes which results in extreme reliance on key individuals for the success of the project and the heuristics they have developed. That leads to a quick-fix approach to all software maintenance projects, which does not consider ripple effects of its actions, which is always in fire-fighting mode without regards to other areas that may get affected. There is no time for documentation or following guidelines.

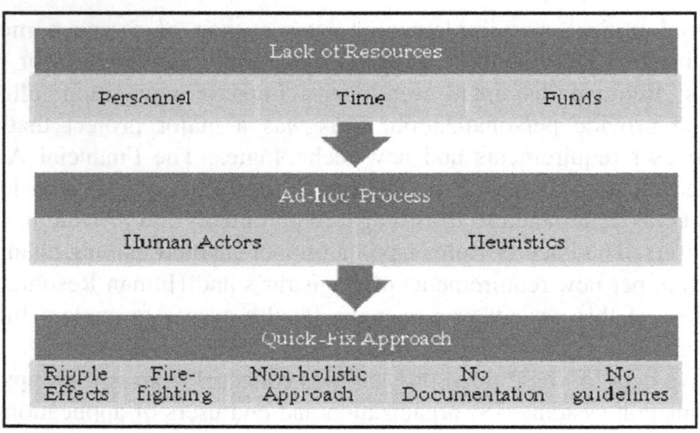

Fig. 1 Why is Software Maintenance so Challenging?

As discussed above our analysis revealed that the software maintenance processes within this organization are essentially ad-hoc. However, we also found evidence that in the absence of formal software maintenance guidelines the organization has developed a pattern where they rely on their own rules gleaned from the collective experiences over the years. These rules developed from the ad-hoc processes and unwritten guidelines used in the maintenance projects represent heuristics that are followed in any new projects that are initiated.

The term heuristics was coined by Simon who defined them as "adaptive strategies undertaken by humans when their capacity is limited for information processing" (Simon 1957). Existing research shows that people use heuristics to make sense of their decisions. For example studies in political science reveal that citizens often employ different heuristics to make sense of politics (Bray and Sniderman 1985). If a politician is Republican, then voters readily infer that s/he is for low taxes, strong defense, against government intervention in the economy, etc. Heuristics are used often and are associated with higher quality decisions. If heuristics did not work at least some of the time, they would not be developed and utilized (Lau and Redlawsk 2001). Further, in organizational settings heuristics often represent accumulated wisdom of experts or experienced resources. As Sniderman, Brody and Tetlock mentioned, "the comparative advantage of experts is not that they have a stupendous amount of knowledge, but that they know how to get the most out of the knowledge they possess" (Sniderman et al. 1991). In other words, not only do experts or experienced resources employ certain cognitive heuristics, but they are very much likely to employ them appropriately (Lau and Redlawsk 2001).

During our analysis of the operations within the organization we found evidence of three different kinds of heuristics. Categorizations of these heuristics are based on the analysis of qualitative data collected via interviews. The criteria to distinguish these types are as follows. The first is organizational heuristics that has a broader span and is informally implemented by management across the organization. The second is individual heuristics. It represents rules adopted by experienced individuals within the organization. These rules tend to be different from one individual to another. The third group of heuristics is related to the size and phase of each project. We provide further details about these below in Tables 2, 3 and 4, respectively, and discuss them in the subsequent sections.

3.1 Organizational Heuristics

The organizational heuristics that were identified represented rules used within the organization for skill development, resource augmentation and coordination strategies. The first five rules deal with overcoming challenges of maintaining skills and staff development. These are important as small organizations, despite being short on resources, have to be prepared and must deliver solutions. Rules 6, 7 and 8 deal with strategies of getting help from outside. Rules 9, 10 and 11 suggest ways in which office arrangements, hiring and meetings can be utilized to better assist small organizations. Details are provided in Table 2.

Table 2 Organizational Heuristics' Application to Projects

	Organizational Heuristics	Portal	Financial	Faculty
1.	If missing key skills, adopt formalized training	✓		
2.	If a project needs a backup person and junior staff is available, pair with senior member for cross training	✓	✓	
3.	If time allows and topics are related to project's subject matter, send your staff to local conferences	✓	✓	✓
4.	Send webinar info. & let staff decide if they want to attend	✓	✓	✓
5.	Involve senior staff in all projects – especially in design	✓	✓	✓
6.	Use outside consultants for high profile projects	✓	✓	
7.	For mission critical needs, acquire formal product support	✓		
8.	For technical problems search a solution on Google first	✓	✓	✓
9.	Hire individuals with entrepreneurial traits	✓	✓	✓
10.	Enable office arrangement that enable communication and mutual support. Cubicles are better than walled offices	✓	✓	✓
11.	Utilize meetings to manage individuals and get tasks done	✓	✓	

3.1.1 Formal Training

Due to financial constraints, small organizations don't have a training budget. They, however, have to be at the cutting edge and sometimes need to avail formal training. The cost for sending one person to such training is very high (thousands of dollars). Selection of training option depends on impending large project that requires certain expertise. Thus, the rule is if a high profile project is looming and certain skills are short, adopt formal training.

3.1.2 Cross Training

Small organizations often have low skilled individuals who are brought up-to-speed by pairing them with experienced persons. It, however, can take away time from experienced staff. One senior developer stated:

"I have a lot on my plate. If I am expected to train juniors, that eats into my time. It gets frustrating, especially if the person I am training is not very motivated."

Thus, the rule is if a project needs a backup person and junior staff is available; pair them with senior members for cross training.

3.1.3 Attending Conferences

Due to exorbitant costs, frequent formal training is often out of the reach of small organizations. Therefore conferences especially local ones provide an excellent mean of staying current. They allow networking with peers, sharing of ideas, codes and design techniques. Thus the rule is if time allows send your staff to local conferences.

3.1.4 Webinars

Webinars is an important resource for technical knowledge. Management encourages its staff to pursue these. However, finding and recommending the right webinar can be time consuming. Thus the rule is to send webinar information to staff and let them decide if they are worth attending.

3.1.5 Involve Senior Staff

Involving senior staff allows the organizations to utilize their wisdom and experience. One developer mentioned:

"Must involve experienced staff in the beginning of the project…they are like the brain of the system that tell you where to start. Without them, you may be lost."

As senior staff members are few in numbers, they should be allocated more to design work and less to low-level coding which should be assigned to junior staff. One caveat is to keep skills of senior staff current. One senior developer stated:

"I want to be involved in coding otherwise this stuff disappears from brain."

Furthermore, as small organizations do not have good documentation systems, senior staff is an excellent source of information. They have institutional memory.

3.1.6 Involving Consultants

Involve consultants when the project has a high profile and needs to be completed in a specified time period. In the current site, it was learned when consultants were involved, a sense of urgency came into play. One developer stated:

"Everyone knows that consultants charge by the hour and we better provide them with what they need to get the job done. They bring focus."

Thus the rule is to bring consultants in for projects that are high profile and are needed in a specified time period.

3.1.7 Formal Support

Oftentimes solutions cannot be easily found through other means and one has to rely on vendor support. While small organizations don't have financial means to acquire such support for all applications, it should consider purchasing support for mission critical applications.

3.1.8 Google Search

Current search engines and Internet provide vast and free resources to organizations. Organizations encourage staff to look up solutions to problems and ideas on the web. One developer stated:

"When confronting a problem my manager often asks me: have you Googled it?"

Thus the rule is when investigating a problem, search for a solution on Google.

3.1.9 Hiring Practices

The staff members of small organization must be willing to learn new technologies, work on multiple projects, be able to handle stress and must be willing work independently without guidelines. These are characteristics of a person who is running his own business. One analyst mentioned:

"Those people who have been entrepreneurs work best in our shop."

Thus the rule is when hiring look for individuals with entrepreneurial background.

3.1.10 Use Meetings to Manage Individuals

Most small organizations, as evidenced in our case, operate under functional management style. The real power lies with functional managers. PMs operate without proper authority due to which they find it difficult to get team members to commit to deadlines and finish tasks. To overcome this hurdle they hold regular status meetings. One analyst stated,

"status meetings are good reminders for us to get tasks done. As people know they have to give status, they are motivated to show results by the meeting time."

Thus the rule is to utilize meetings to get tasks done.

3.1.11 Office Proximity

It was learned that timely communication plays a vital role. Quick and earshot communication is enabled through closed proximity of staff offices and utilization of cubicles. One analyst mentioned:

"Because we sit close to each other, we quickly exchange information. To get answers, I don't want to wait till meetings, back and forth emails or play phone tags. Ours is creative work, if communication is not instant; it's hard to refocus."

Thus the rule is to seat staff close to each other for ease in communication and for quick mutual support.

3.2 *Individual Heuristics*

The vacuum created by formal guidelines is also filled by individual heuristics developed over the years. Below is a list of such rules identified through the case study. The first three rules deal with individuals' efforts in maintaining their skills. Rules 4, 5 and 6 emphasize problem identification and ripple effects of fixes. Rules 7 and 8 pertain to the benefits of involving users in software maintenance. Rules 9, 10 and 11 suggest maintaining documentation, preparing for meetings and utilizing the "Find" tool in software maintenance.

3.2.1 *Adopt Non-formal Methods to Stay Abreast of Changes in the Field*

In small organizations, staff is expected to manage maintenance with little support. If they don't perform, they are replaced. They turn to formal training, but that is often ill fitted to their needs and is out of reach. One analyst mentioned:

"Formal training is a waste of time. They cover only surface areas and don't go deep enough. The best way for me to learn is on the job and on the project."

Thus the rule is to adopt non-formal methods such as on-the-job learning or create self exercises to stay abreast of changes.

Table 3 Individual Heuristics' Application to Projects

	Individual Heuristics	Portal	Financia Aid	Faculty Roles
1.	Adopt non-formal methods to stay current in the field	✓	✓	✓
2.	To ensure continuity, be involved in all phases	✓	✓	✓
3.	Be prepared for future changes and projects	✓	✓	✓
4.	Identify the problem well	✓	✓	✓
5.	Consider ripple effects	✓	✓	
6.	In the absence of documentation, talk to experienced staff	✓	✓	
7.	Prior to the start of a project, make test plans with users	✓	✓	
8.	Identify and involve power users	✓	✓	✓
9.	Maintain a personal documentation system	✓	✓	
10.	Prepare for meetings prior to attending	✓	✓	
11.	Use "Find" in an editor for software maintenance coding		✓	✓

3.2.2 Be Involved in All Phases of a Project

Staff members' involvement in multiple phases of a project led to increased awareness of issues, continuity and better communication. This rule could be categorized as organizational; however, management finds it impractical to assign individuals to all phases. Thus individuals should take initiative, and as possible, be involved in all phases.

3.2.3 Be Prepared for Future Projects

An important rule that emerged was for individuals to be inquisitive and informed of the new projects on the horizon. One developer stated:

"I often learn about new projects over the grapevine. We don't have formal methods of knowing about upcoming projects unless our manager informs us."

Thus the rule is to be on the lookout. Once knowledge is gained of an upcoming project, start acquiring appropriate knowledge, background and training.

3.2.4 Spend More Time on Identification of Problems

Ample time should be spent on identification of problems. One developer stated:

"Sometimes I have spent days trying to figure out a problem and then I make one line of code change and that takes care of the takes care of it."

Thus the rule is to give good amount of time to identification of the problem.

3.2.5 Consider Ripple Effects

Software maintenance changes and effects should be identified and considered prior to implementing them. One developer said:

"I must identify if I am going to fix one problem but break 10 others."

Thus the rule is prior to implementation of any change, consider its ripple effects.

3.2.6 Talk to other Experienced Staff

While confronting a given project, it is best to see if a similar project has been undertaken before. In the absence of documentation, the best resource for this is senior staff. One developer stated:

"Often I would turn my wheels around until I find a solution. It's best that I get some direction from senior staff, especially since we are short on time."

Thus the rule: Every project should be vetted through senior staff.

3.2.7 In Testing Involve Users

In small organizations users should be involved in testing from the beginning of the project. That way software maintenance will be done exactly per users' requirements and with fewer bugs.

3.2.8 Work with Power Users

Power users have the highest interest and motivation in making their application project a success. Power users typically do not waste the time of I.T. resources and often go out of the way to help the project. Thus the rule is to identify and involve power users in software maintenance.

3.2.9 Keep a Personal System of Documentation

In the absence of a central documentation system, individuals should create their own systems so that project information is readily available as needed. One analyst mentioned:

"I have to maintain my own information to be able to maintain my job."

Thus the rule is to maintain personal documentation system.

3.2.10 Be Prepared for Meetings

Before attending a meeting, doing homework is highly recommended. That allows one to gain the most from meetings, contribute to them and also not be bored. As one of our respondents note:

"Before going to a meeting, I spend at least fifteen minutes preparing. That keeps me tuned in during the meeting; otherwise, it is hard to keep up with stuff,"

Thus the rule is to be prepared for meeting prior to attending.

3.2.11 Use of Find

It was learned that the most commonly used tool in maintaining systems is the "Find" tool in code editors. One developer stated,

"While I am asked to work on a software maintenance project, the first thing I do is identify the relevant material by the Find feature of a code editor."

Instead of going through the whole program and understanding it, Find is used to get to the relevant information. In most cases documentation is non-existent; thus the rule in software maintenance coding is to use a "Find" tool.

3.3 Heuristics Regarding Different Stages of Projects

While abovementioned were heuristics that were either organizational or individual in nature, some heuristics were found to be associated with the size component and phase dimension of each project. In the current site, majority of projects are

Table 4 Heuristics' Application to Osborne's Phases (Osborne, 1987)

Osborne Stages	Large / Medium Projects	Small Projects
Initiation	Involve Executive Committee	Manager Initiates
Requirements Gathering	Invite all concerned to a meeting. Involve technical people in requirement and design analysis so nothing is missed	Have the end user send in requirements through email. Involve developers to gather requirements
Development	Allow ample time for requirement gathering. Adopt an iterative development methodology	Use prototyping approach. Develop a sample and improve in iterations. Use one user & one developer
Testing	Involve users in unit testing. Have as few users as possible in testing	Development and testing are intertwined
Implementation	Allow testing to continue after implementation (bugs fixing)	

of medium size and of perfective/adaptive maintenance type. It was learned that as the risk of a project failure increases, multiple layers of management jump in to guide it. That causes some confusion/conflict but keeps the project moving. Listed below are some of the other heuristics learnt from the current study. They are mapped along the phases of Osborne's methodology of software maintenance.

4 Discussion and Future Research Plans

The initial results of our case analysis indicate software maintenance operations in small organizations are often carried out using informal ad-hoc processes. However such ad-hoc processes are reinforced by various levels of heuristics. We identified heuristics related to the organizational resourcing and training, individual approach to tasks as well as those related to different stages of a project. Therefore an outcome of this research seems to be the salience of heuristics in small organizations. They represent an intermediate framework that compensates for the lack of formal procedures, and emerge as a response to resource straitened context of small organizations. While heuristics seem to be a useful mechanism for small organizations, there are certain caveats to their use that need to be noted.

Heuristics typically represent wisdom accumulated from experience. Though the experience is real and the lessons captured are valuable, different experiences easily can lead to contradictory guidelines. For instance, there are software engineers who firmly believe that C++ (a statically typed language) is too inflexible for serious system development. On the other hand, there are those who believe that the error-prevention properties of such statically typed languages are required for serious system development (Stacy and Macmillian 1995). These types of perceptions in formulation of heuristics need investigation in the context of software maintenance. For this to be done it is important to understand how individuals formulate heuristics. Tversky and Kahneman (1973) noted that "in making predictions and judgments under uncertainty, people do not appear to follow the calculus of chance or the statistical theory of prediction. Instead, they rely on a limited number of heuristics which sometimes yield reasonable judgments and sometimes lead to severe and systematic errors" (Kahneman and Tversky 1973).

Kahneman and Tversky have further defined three cognitive heuristics and biases that can emerge: representativeness, availability, and anchoring-and-adjustment. Representativeness refers to making an uncertainty judgment on the basis of "the degree to which it similar to its parent population and reflects the salient features of the process from which it is generated". Availability is used to estimate "frequency or probability by the ease with which instances or associations come to mind". Availability is influenced by imaginability, familiarity, and vividness, and is triggered by stereotypical thinking. Anchoring-and-adjustment involves "starting from an initial value that is adjusted to yield the final answer. The initial value, or starting point, may be suggested by the formulation of the problem, or it may be the result of a partial computation. In either case, adjustments are typically insufficient" (Kahneman and Tversky 1979). In other words, when people are given a problem for which there exists a correct solution and an initial estimate of its solution, they tend to provide final estimates close to the initial estimate when asked to solve the problem (Parsons and Saunders 2004). This is termed anchoring. The anchoring heuristic helps humans simplify problem solving in a complex situation without conscious effort. However, since they are not aware they are using the heuristic, people often fail to make adequate modifications to an initial solution and produce estimates that are severely flawed with respect to the correct solution. This is termed an adjustment bias (Shanteau 1989).

This discussion shows that while heuristics could be potentially useful tools for software maintainers within small organizations, they are susceptible to cognitive biases in their formulation and use. These biases can affect the way in which software maintenance artifacts are reused, potentially impeding or enhancing the successful reuse of code and design. It is important therefore to understand how heuristics are formulated in the software maintenance context and categorize them to facilitate further in depth study. Kahneman and Tversky's taxonomy of cognitive heuristics and biases perhaps form a good foundation for such investigation (Kahneman and Tversky 1979).

It is also perhaps critical to try and understand how the biases in heuristic formulation can be compensated. Specifically there is a need to investigate what can be done to reverse biases. Existing research provides four recommendations that can alleviate these biases. Firstly, there is a need to frame problems to signify relevant information that might be otherwise left out. Secondly, use specific education/training to raise awareness of common biases. In general, however, cognitive biases do not disappear just because people know about them. Thirdly, software maintainers should understand that their impressions of an application may be biased, and should undertake empirical investigation whenever possible. If they need to know how often a certain idiom occurs in code, it will often be wise to use an automated tool to find out. If they need to know where an application is spending most of its time, profiling should be utilized. Lastly, they should attempt to seek disconfirmatory information when applying heuristics. For example when testing and debugging the application, engineers should actively seek disconfirmatory information (Stacy and Macmillian 1995). This would enable individuals to continuously evaluate the applicability of the adopted heuristics to the context of use.

In conclusion, an important finding of our research is the identification of the fact that small organizations often resort to heuristics to compensate for the lack of formal processes. However while heuristics represent a viable mean to structure software maintenance operations within an organization, the discussion above indicates that there are certain problems associated with their use and formulation and they should therefore be used with caution. We feel that our research provides some initial insights into the performance of software maintenance in small organizations. However this represents preliminary research and future research needs to examine the formulation and biases in heuristic formulation within software maintenance to enable more efficient use of these mechanisms within small organizations.

References

1. Anquetil, N., De Oliveira, K.M., De Sousa, K.D., Batista Dias, M.G.: Software maintenance seen as a knowledge management issue. Information & Softw. Technology 49(5), 515–529 (2007)
2. April, A., Hayes, J.H., Abran, A., Dumke, R.: Software maintenance maturity model (SM^{mm}): The software maintenance process model. J. of Softw. Maintenance & Evolution: Research & Practice 17(3), 197–223 (2005)
3. Benbasat, I., Goldstein, D.K., Mead, M.: The case research strategy in Information Systems. MIS Q. 11, 369–386 (1987)
4. Bray, H.E., Sniderman, P.M.: Attitude attribution: A group basis for political reasoning. American Political Science Rev. 79, 1061–1078 (1985)
5. Brodman, J.G., Johnson, D.L.: What small businesses and small organizations say about the CMM. In: Proceedings of 16th International Conference on Software Engineering (ICSE 1994), pp. 331–340. IEEE Computer Society Press, New York (1994)
6. Darke, P., Shanks, G., Broadbent, M.: Successfully completing case study research:combining rigor relevance and pragmatism. Information Systems J. 8, 273–289 (1998)
7. Garcia, S.: Thoughts on applying CMMI in small settings. Presented in Carnegie Mellon Software Engineering Institute (2005), `http://www.sei.cmu.edu/cmmi/adoption/pdf/garciathoughts.pdf`

8. Grubb, P., Takang, A.A.: Software maintenance concepts and practice, 2nd edn. World Scientific Publishing, Singapore (2003)
9. Hofer, C.: Software development in Austria: Results of an empirical study among small and very small enterprises. In: Proceedings Euromicro Conference (2002)
10. IEEE, Std 1044, IEEE standard classification for software anomalies, IEEE (1993)
11. Kahneman, D., Tversky, A.: On the psychology of prediction. Psychological Rev., 237–251 (1973)
12. Kahneman, D., Tversky, A.: Intuitive prediction: Biases and corrective procedures. TIMS Studies in Management Sciences, 313–327 (1979)
13. Lau, R.R., Redlawsk, D.P.: Advantages and disadvantages of cognitive heuristics in political decision making. American J. of Political Science 45(4), 951 (2001)
14. Lientz, B.P., Swanson, E.B.: Software Maintenance Management. Addison Wesley, Reading (1980)
15. Orlikowski, W., Baroudi, J.: Studying information technology in organizations: Research approaches and assumptions. Information Systems Research 2.1, 1–28 (1991)
16. Osborne, W.M.: Building and Sustaining Software Maintainability. In: Proceedings of Conference on Software Maintenance, pp. 13–23 (1987)
17. Parsons, J., Saunders, C.: Cognitive heuristics in Software Engineering: Applying and extending anchoring and adjustment to artifact reuse. IEEE Transactions on Softw. Engineering 30(12), 873–888 (2004)
18. Pigoski, T.M.: Practical software maintenance: Best practice for managing your software investment, pp. 29–36, 117–138. Wiley, New York (1997)
19. Shanteau, J.: Cognitive heuristics and biases in behavioral auditing: Review, comments and observations. Accounting, Organizations & Society 14(1/2), 165–177 (1989)
20. Simon, H.A.: Models of Man: Social and Rational. Wiley, New York (1957)
21. Sniderman, P.M., Brody, R.A., Tetlock, P.E.: Reasoning and choice: Explorations in Political Psychology. Cambridge University Press, New York
22. Software CMM. Mid-Year Update by Software Engineering Institute (2005), `http://www.sei.cmu.edu/cmmi/casestudies/profiles/pdfs/upload/2005sepSwCMM.pdf`
23. Software Industry Statistics for 1991-2005, Enterprise Ireland (2006), `http://www.nsd.ie/htm/ssii/stat.htm`
24. Stacy, W., Macmillian, J.: Cognitive bias in Software Engineering. Communications of the ACM 38(6), 57–63 (1995)
25. Staples, M., Niazi, M., Jeffery, R., Abrahams, A., Byatt, P., Murphy, R.: An exploratory study of why organizations do not adopt CMMI. J. of Systems & Softw. 80(6), 883–895, 13 (2007)
26. Takang, A.A., Grubb, P.A.: Software maintenance concepts and practice. Thompson Computer Press, London (1996)
27. Walsham, G.: Doing Interpretive Research. European J. of Information Systems 15, 320–330 (2006)
28. Yin, R.: Case study research design and methods, 2nd edn., vol. 9. Sage Publications, Thousand Oaks (1994)

Are We Relying Too Much on Forensics Tools?

Hui Liu, Shiva Azadegan, Wei Yu, Subrata Acharya, and Ali Sistani

Abstract. Cell phones are among the most common types of technologies present today and have become an integral part of our daily activities. The latest statistics indicate that currently there are over five billion mobile subscribers are in the world and increasingly cell phones are used in criminal activities and confiscated at the crime scenes. Data extracted from these phones are presented as evidence in the court, which has made digital forensics a critical part of law enforcement and legal systems in the world. A number of forensics tools have been developed aiming at extracting and acquiring the ever-increasing amount of data stored in the cell phones; however, one of the main challenges facing the forensics community is to determine the validity, reliability and effectiveness of these tools. To address this issue, we present the performance evaluation of several market-leading forensics tools in the following two ways: the first approach is based on a set of evaluation standards provided by National Institute of Standards and Technology (NIST), and the second approach is a simple and effective anti-forensics technique to measure the resilience of the tools.

Keywords: Cell phone forensics, Android, Smart phone, Cell phone forensics tool, Anti-forensics.

1 Introduction

Cell phones usage has been growing for the past several years and this trend is expected to continue for the foreseeable future. Smart phones with increasing storage capacity and functionality have outnumbered PCs and are becoming the media of choice for personal use, especially among the younger generations. Large amount of information from phone book to photo albums and videos, to emails and text messages, to financial records and GPS records, are stored in these phones. Smartphones are highly capable of internetworking, especially through wireless connections such as WiFi, 3G and 4G. It is undeniable that smartphones are becoming a more and more important type of terminal device of the entire Internet.

Hui Liu · Shiva Azadegan · Wei Yu · Subrata Acharya ·Ali Sistani
Department of Computer & Information Sciences, Towson University, Towson, MD 21252
e-mails: `janetliuhui@gmail.com`,
`{azadegan,wyu,sacharya}@towson.edu`, `masistani@yahoo.com`

R. Lee (Ed.): Software Eng. Research, Management & Appl. 2011, SCI 377, pp. 145–156.
springerlink.com

As the use of cell phones grows in popularity, there has been an increase in their involvement in digital crimes and digital forensics. The latest statistics indicate that over five billion cell subscribers are in the world [1–3] and increasingly cell phones are used in criminal activities and confiscated at the crime scenes (e.g., cops warn of more cyber crimes with the launch of 3G services [4]). Data extracted from these devices are presented as evidence in the court that has made cell phone forensics a critical part of law enforcement and legal systems in the world.

Because of the pervasive use of cell phones, recovering evidence from cell phones become an integrated part of digital forensics and also poses great challenges. In particular, cell phone forensics, a relatively a new field, is largely driven by practitioners and industry. Some of the challenges associated with the analysis of cell phones are due to the rapid changes in technology. These devices are continually undergoing changes as existing technologies improve and new technologies are introduced. Forensic specialists today operate within what is called the forensic tool spiral. New versions of forensic tools are created regularly to include the new features and functionality of the fast evolving phones. Cell phones with rich capacities have various peripherals that make data acquisition a daunting job. [5–7]. Since these tools are routinely used by law enforcement and legal organizations to assist in the investigation of crimes and used to create critical evidence used in the court, it is of outmost importance that the results can be validated.

Due to the lack of standard validation procedures for forensics tools, the researchers at NIST working on the Computer Forensics Tool Testing (CFTT) project have developed a set of assertions and test cases that can be used for evaluating forensic tools [8,9]. During the first phase of this project, we conducted a thorough evaluation of the five most commonly used forensic tools using the evaluation standards provided by NIST. Through this process we assessed the reliability and effectiveness of each tool in extracting logical data from an Android-based phone and ensured that the tools, under normal circumstances, extract data successfully from a given phone. In the second phase of the project, thinking like an adversary, we investigated the resilience of these tools to an anti-forensics technique. Its basic idea is to enable phone protect the privacy of its data and stop the data extraction process when it recognizes that it is connected to a forensics tool.

The remainder of the paper is organized as follows. In Section 2, we present a brief overview of the related work. In Section 3, we systematically introduce the forensics tools used in this project and their evaluation results based on standards provided by NIST. In Section 4, we discuss the implementation of a simple and effective anti- forensics Android application and the evaluation results. In Section 5, we present the conclusion and outline our future research directions.

2 Related Work

Cell phone forensics aims at acquiring and analyzing data in cellular phones. Forensics tools for cell phones are quite different from those for personal computers.

There have been a number of efforts for evaluating forensics tools and developing forensics techniques for cell phones in the past. For example, Curran *et al.* presented mobile phone forensic analysis, what it means, who avails of it and the software tools used [10]. Somasheker *et al.* presented the overview of tools available for examining PDAs and cell phones [11]. Kim *et al.* presented the technique to collect cell phone flash memory data using the logical method to acquire files and directories from the file systems of the cell phone flash memory [12]. Ting *et al.* investigated the dynamic behavior of the mobile phone's volatile memory and presented an automated system to perform a live memory forensic analysis for mobile phones [13]. Mokhonoana *et al.* presented the forensics technique, which uses an on-phone forensic tool to collect the contents of the device and store it on removable storage [14]. Their approach requires less equipment and can retrieve the volatile information that resides on the phone such as running processes. Connor presented the forensics techniques based on the analysis of call detail records (CDR) and the corresponding tower-antenna pairs, which can provide useful information as evidence in a criminal trial [15]. Different from those existing research, our work is a thorough evaluation of the five most commonly used forensic tools based on the evaluation standards provided by NIST published recently.

Our work is also related to anti-forensics. Distefano *et al.* discussed the classification of the classification of the anti-forensics techniques (e.g., destroying evidence, hiding evidence, eliminating evidence sources, and counterfeiting evidence) [16]. Garfinkel also categorized the traditional anti-forensics technique. [17]. Different from the exist- ing research efforts, our work is the first one, which enables the protection of privacy of phone's data and stop the data extraction process when the phone recognizes that it is connected to a forensics tool.

3 Forensic Tools and Evaluation

During the first phase of this project, we have been using the evaluation standards pro- vided by NIST [8, 9] to evaluate the five market-leading commercial forensics tools discussed below. In November 2010, researchers at NIST also provided a thorough evaluation of four of these tools [18–21]. In this section, we first briefly describe these tools and then present the summary of our evaluation results.

3.1 Forensic Tools

The five most commonly used forensic tools are describe below.

- Cellebrite [22] is a tool for mobile phone, smart-phone, and PDA forensics. As of September 2010, the tool is compatible with over 2,500 mobile phones (including GSM, TDMA, CDMA, iDEN). The tool has an integrated SIM reader with wireless connection options and supports all known cellular device interfaces, including Serial, USB, IR, and Bluetooth. The tool also supports native Apple iPOD Touch, and Apple iPHONE extraction on both 2G and 3G versions, as well as iOS4. Cellebrite is simple to use and enables storage of hundreds of phonebooks and content items onto a SD card or USB flash drive. Extractions can then be brought back to the forensic lab for review and verification using the reporting/analysis tool. Subject data can be retrieved either via logical or via physical extraction methods.

– Paraben Device Seizure [23] is an advanced forensic acquisition and analysis tool for examining cell phones, PDAs, and GPS devices. The tool is able to acquire both logical and physical data. It contains a report generation tool that allows for the convenient presentation of data. The tool is able to generate reports using the report wizard in csv, html, text or xls formats. The tool is designed to recover the deleted data and retrieve physical data from some devices and has a fairly simple user interface. It provides features to convert GPS data points to be read directly into Google Earth so investigators can quickly and easily visualizes where these GPS locations map.
– XRY [24] is one of the best-dedicated mobile device forensic tools developed by Micro Systemation (MSAB). 'XRY Complete' is a package containing both soft- ware and hardware to allow both logical and physical analysis of mobile devices. The unified logical/physical extraction wizard in XRY and the resulting reports help to show the examiner the full contents of the device in a clean and professional manner. The tool is able to connect to the cell phone via IR, Bluetooth or cable interfaces.

After establishing the connectivity, the phone model is then identified with a corresponding picture of the phone, name of the manufacturer, phone model, its serial number (IMEI), Subscriber ID (IMSI), manufacturer code, device clock, and the PC clock. The retrieved information report is easy to read, with the data explorer on the left-hand side, including the Summary, Case Data, General Information, Contacts, Calls, SMS, MMS, Pictures, Videos, Audio, Files, and Log information. XRY re- ports can be exported to Microsoft word, xls, xml, OpenOffice, and txt formats. The tool is user friendly and performs the best amongst all our test forensics tools on our evaluation phone - the Motorola Droid 2 A955. Additionally, it is the only tool that can detect the phone correctly with its exact phone type and provides a very detailed device manual for analysis and evaluation.

– EnCase [25] is a family of all-in-one computer forensics suites sold by Guidance Software. Users can create scripts, called EnScripts, to automate tasks. In particular, the EnCase Neutrino tool captures and processes mobile device data for analysis and review alongside other collected data. The solution features hardware support and parsing capabilities for the most common mobile device manufacturers, including Nokia, Motorola, Samsung, Siemens, LG, Palm, BlackBerry, HTC, UTStar- com, Sony Ericsson, etc. Investigators can collect potentially relevant ESI stored in the mobile device settings. The tool features a specially produced WaveShield wireless signal-blocking bag, which can help users view and capture cell phone data in a forensically sound manner even within close proximity to cell towers. The tool can correlate and analyze data from multiple devices with collected computer evidence and aid in sharing Logical Evidence Files collected and preserved from mobile devices with other EnCase Examiners. Note that as of April 2011, Guidance Software has discontinued Encase Neutrino and changed its focus towards the development efforts on expanding support of Smartphone and tablet devices via the Encase Smartphone Examiner toolkit.

- The SecureView [26] tool provides various smart features for cell phone forensics such as: svSmart (ability for streamlining and presetting conditions to attain evidence in the field quickly); svPin (ability for unlocking CDMA cell phone pass- words); svLoader (ability to download, analyze, verify and validate other sources and aid in creating csv files and upload, back up files from RIM and iPhone); and svReviewer (ability to share data without multiple licenses). The tool contains a SIM card reader to extract data from the SIM cards of GSM phones. The tool acquires data via cable, Bluetooth or IR interfaces. Reports are then generated in a print ready format.

  It provides a friendly interface and strong phone support for over 2000 phones. The main interface includes the phone, memory card, SIM card, Svprobe and report. Svprobeis primarily the interface for data analyzing. The tool provides both manual and automatic connection options. The tool also provides the option of generating a detail report of data and output in a pdf format. Additionally, the report can be merged with the import report from other forensic tools by using Svprobe. The tool provides five connector options, which are the USB cable, serial cable, IR, Bluetooth and SIM card reader. One of the advantages for data acquisition with the tool is that it can use cables from other tools (e.g., Cellebrite).

3.2 Evaluation

There are three primary methods of acquiring data using the above forensic tools:

- *Manual Acquisition.* It implies that the user interface can be utilized to investigate the content of the memory. This method has the advantage that the operating sys- tem makes the transformation of raw data into human interpretable format. The disadvantage of this method is that only data visible to the operating system can be recovered and that all data are only available in form of pictures.
- *Physical acquisition.* It implies a bit-by-bit copy of an entire physical storage. This acquisition method has the advantage of allowing deleted files to be examined. However, it is harder to achieve because the device vendors needs to secure against arbitrary reading of memory so that a device may be locked to a certain operator.
- *Logical acquisition.* It implies a bit-by-bit copy of logical storage objects that reside on a logical store. This method of acquisition has the advantage that system data structures are easier for a tool to extract and organize. It acquires information from the device using the interface for synchronizing the contents of the phone with the analyzing device (e.g., PC). This method

usually does not produce any deleted information, due to it normally being removed from the file system of the phone.

In accordance with the method used by the NIST evaluations, in this project we mainly focus on the method of Logical acquisition of data and their subsequent analysis for our evaluation and the development of our proposed anti-forensic application.

The following nine evaluation categories were identified by CFTT project as the requirement for Smart phone tool evaluation [8, 9]. We have summarized our evaluation results for each of these categories listed below.

- *Connectivity:* In regards to the connectivity support by the device through suit- able interface - Cellebrite, Paraben and XRY recognized the Android phone; both the Motorola and Android phones were recognized by SecureView; and HTC was recognized but failed for Neutrino. In the test for connectivity attempt to non- supported devices - a wide range of devices were supported by Cellebrite; a new driver was required by SecureView; and there was a no show in the phone list as found for Neutrino, Pareben and XRY. Finally, on a connectivity interrupt test - Cellebrite displayed a warning message; Packet failed message was displayed for SecureView and Neutrino; and Packet failed received message was displayed for Paraben and XRY.

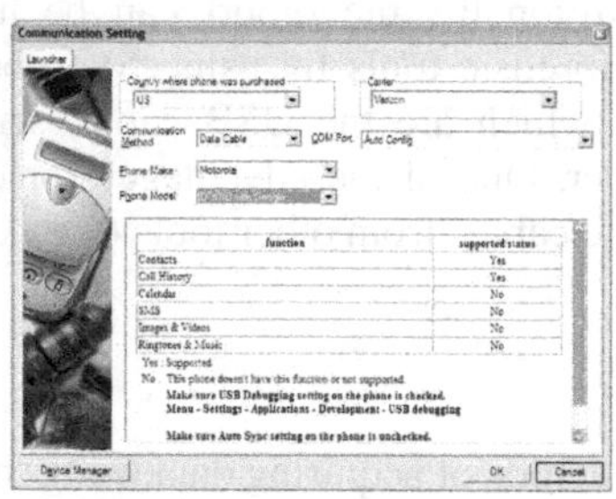

Fig. 1 Initial Communication Setting for SecureView

- *Data Acquisition:* All the forensic tools presented the data acquired in a usable format via the generation of reports. With regards to the present subscriber related information in a usable format for target devices - the IMSI was retrieved for Cellebrite; the IMSI message was entered by examiner for SecureView; the IMSI, the Network ID and phone number was generated by Neutrino; Paraben was able to obtain the phone number and subscriber ID; and IMSI and MSISDN was generated for XRY. With regards to the presentation of equipment related information in usable format for target device - IMEI was retrieved for Cellebrite; examiner entered the message for SecureView; Neutrino generated a serial number; Paraben presented the device ID and phone number; and the IMEI was generated for XRY. The SPN was not detected in case of Cellebrite; for SecureView and Paraben, it was not applicable as there was no SIM card and the CDMA was without SIM respectively; network code 'SPN=?' was displayed for Neutrino;

and only XRY presented the SPN in usable format. With regards to presenting the ICCID in usable format for SIM-Cellebrite, Neutrino and XRY did acquire the ICCID; and there was no applicability for SecureView and Paraben. Similarly, for the test to present IMSI in usable format for SIM-ICCID was acquired for Cellebrite, Neutrino and XRY; and there was no applicability for SecureView and Paraben.

– *Location:* With regards to the acquisition of location data - Cellebrite, Neutrino and XRY did acquire the data; and the test was not applicable for SecureView and Neutrino. For the test for acquisition of GPRS location data - no GPRSLOCI was acquired for Cellebrite; no routing area location (RAI-LAC) was detected for Neutrino and XRY; and the test was not applicable for SecureView and Paraben.
– *Selection Option:* For the selection option for Acquire all supported device objects
- the test was a success for all devices except for Cellebrite. With regards to the Ac- quire all supported device objects by individually selecting each supported object - the test was a success for Neutrino, Paraben and XRY; the message was displayed to the user for SecureView; but it was not supported for Cellebrite. Finally, for the selection option of Selecting individual data objects - all data objects was acquired instead of individual objects for Cellebrite; and the test was a success for Secure- View, Neutrino, Paraben and XRY.

Paraben
host::.
device::.
OPEN
sync:.
RECV…./system/build.prop
shell:pm uninstall com.paraben.service.
STAT …./data/local/tmp
SEND(…/data/local/tmp/AndroidService.apk, 61951
shell:pm install /data/local/tmp/AndroidService.apk
pkg: /data/local/tmp/AndroidService.apk
Success.
shell:rm /data/local/tmp/AndroidService.apk
shell:am broadcast –a StartSeizureService.
Broadcasting: Intent {act=StartSeizureService }. Broadcast completed: result=0.
tcp:5000.
<request>. <id>Test</id>.</request>. <request>. <id>Contacts</id>.</request>.
<Test/> <progress>. <position>0</position>. <count>5</count>.</progress>

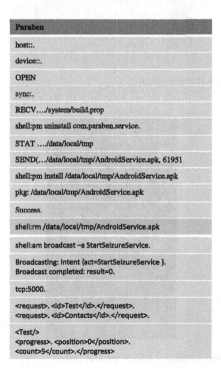

Fig. 2 Pattern of Initial Activities Generated by USBLyzer for Paraben Forensic Tool

- *Authentication:* For the evaluation of SIM password protection - Cellebrite reported no protection; no SIM card was reported for SecureView; and the test result was inapplicable for the other forensic tools. Additionally, the test for the count of PIN and PUK attempts was inapplicable for all tools.
- *Stand-Alone Acquisition:* In regards to the evaluation of stand-alone acquisition on supported stand-alone audio, video and graphic files acquisition - the test was not applicable for Cellebrite; the test was not supported for SecureView; the test failed for both Neutrino and Paraben; and it was not tested for audio and succeeded for video and graphic files for XRY application. Also in regards to the stand-alone acquisition of internal memory with SIM - the test was not applicable for both Cellebrite and SecureView; and there was no stand-alone acquisition for all other tools.
- *Hashing:* For the evaluation of the support for hashing for individual objects - the test was not applicable or had no support for all devices except for XRY application.

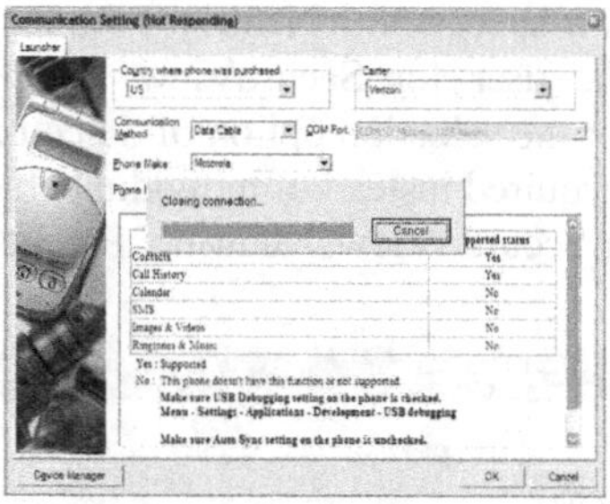

Fig. 3 SecureView Fails to Retrieve Data

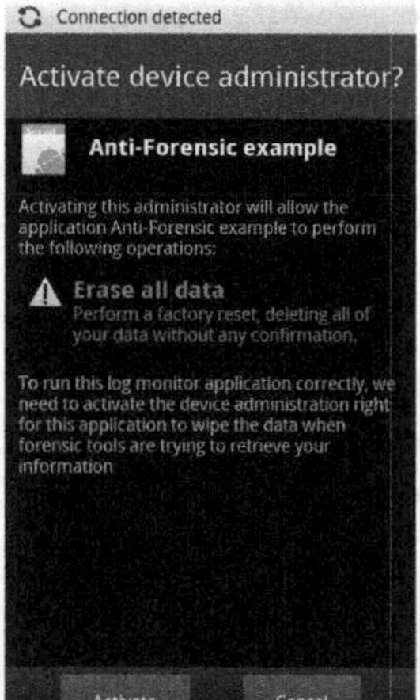

Fig. 4 Authorization of the Application to Wipe Out Data

– *Reporting GPS data acquisition:* In regards to the evaluation of GPS data acquisition - Cellebrite was able to acquire the GPS data for the Android phone; but there was no support for all other tools.
– *Modifications:* Finally, for the modification of case files by third party - our evaluation concluded that the files could not be altered by Cellebrite; SecureView was able to store the files in xls format; XRY was able to edit the files using notepad but could not reopen it further for modification/analysis; and the files could not be reopened for both Neutrino and Paraben tools.

4 An Anti-Forensic Approach

During the second phase of this project, we started thinking like a malicious adversary and investigated ways on how one can compromise the availability of data stored in a phone. Note that Harris [27] defines Anti-forensics (AF) as "any attempt to compromise the availability or usefulness of evidence in the forensic process". In the following, we present a simple and effective Android anti-forensics application, which destroys all data from the phone, as soon as it detects that the phone has been connected to a forensics tool, and makes it unusable during investigative process.

The basic idea of our approach is described as follows: We started our study by monitoring the communication between the tools and the cell phone. We used USBlyzer [28] for this purpose, which is the protocol analyzer software and it captures, decodes and displays communication of forensics tools transmitted through USB device stack.

The basic workflow is shown in Figure 2 listed below. First, the tool sends the application package to /data/local/tmp/ folder. After installing the application, it deletes the package file from the temporary folder and starts the application, which retrieves and transfers the data in the phone to the tool. After the completion of data retrieval the tool uninstalls the application. Comparing the timestamps of these activities, we noticed that there was a multi-second gap between the start of sending the file and start of retrieving information. Our proposed AF application will take advantage of this time gap and as soon as it recognized the initial tool's communication signature, it completely wipes out the data stored on the phone.

To avoid the overhead of running and monitoring all the USB transactions, we decided to monitor the Android logs, which also capture all the activities of the forensics tools. The application uses logcat command to read the log messages line by line and look for the initial connection of the forensic tool. This application that runs as a back- ground service on the phone does not take up much resource. Figure 1 shows Initial Communication Setting for SecureView.

After detection of the forensics tool, the anti-forensics application has to manipulate the information on the android phone before they are retrieved. One option is to alter the sensitive data. The other option is to wipe the device's data (that is, restore the device to its factory defaults). For security concern, we choose later option - to wipe the device's data.

After clicking OK to start to retrieve information from the phone, SecureView starts to install application to the phone. As soon as the phone detects the connection, it shuts itself off and wipes out the data stored in the phone. As shown in Figure 3, SecureView becomes dead while trying to retrieve information from the phone.

We accomplished this task through Android Device Administration API. We first call the DEVICE_ADMIN_ENABLED action that a DeviceAdminReceiver subclass must handle to be allowed to manage a device. This action only needs to be called once for the application. We declared ¡wipe-data¿ as the security policy used in meta-data. Once the user hits the activate button to authorize the application the permission to wipe data, this application will run in the background and wipe the device's data through DevicePolicyManager method wipeData() once it detects the forensics tools' signature. Figure 4 shows the authorization of the application to wipe out data.

The last step is to configure the application to auto-start after boot up; otherwise the application is useless. We used broadcast receiver to implement it. We add the permission RECEIVE_BOOT_COMPLETED to this anti-forensics application to allow it to receive the ACTION_BOOT_COMPLETED that is broadcast after the system finishes booting. Once the broadcast receiver receives the ACTION_BOOT_COMPLETED, it will start the background service to monitor the log.

5 Final Remarks

Anti-forensics is a relatively new discipline and is attracting great attention due to the widespread use of digital devices, particularly, mobile phones and the importance of digital forensics to law enforcement and legal communities. In this paper, we first reviewed and systematically evaluated the performance of several market-leading forensics tools. We then demonstrated how a simple and effective anti-forensics Android application could compromise the availability of the data stored on the phone, and block forensics tools from accessing the data.

We are currently working on the new version of this application with more subtle approaches in deleting the data to carry out anti-forensics. For example, the application only deletes the sensitive data marked as such by its owner from the phone and makes it much harder for the detectives to notice that the data was compromised. Possible countermeasures to anti-forensic techniques are also in the scope of the investigation.

References

1. Group, A.M.: Wireless Network Traffic 2008 2015: Forecasts and Analysis (October 2008), http://www.researchandmarkets.com/reports/660766/
2. Five Billion Cell Users in 2010 (October 2010), http://www.dailywireless.org/2010/02/16/5-billion-cell-users-in-2010/

3. Worldwide Mobile Phone Sales Declined 8.6 Per Cent and Smartphones Grew 12.7 Per Cent in First Quarter of 2009 (May 2009), http://www.gartner.com/it/page.jsp?id=985912
4. Cops warn of more cyber crimes with the launch of 3G services. http://bx.businessweek.com/mobile-tv/view?url=http%3A%2F%2Fc.moreover.com%2Fclick%2Fhere.pl%3Fr4546328679%26f%3D9791
5. Casey, E.: Addressing limitations in mobile device tool. In: Proceedings of the First Annual ACM Northeast Digital Forensics Exchange (2009)
6. Casey, E.: Addressing Limitations in Mobile Device Tool (July 2009), https://blogs.sans.org/computer-forensics/category/computer-forensics/mobile-device-forensics/
7. Casey, E.: Common Pitfalls of Forensic Processing of Blackberry Mobile Devices (June 2009), https://blogs.sans.org/computer-forensics/category/computer-forensics/mobile-device-forensics/
8. NIST, NIST - Computer Forensics Tool Testing (CFTT) Project (2010), http://www.cftt.nist.gov/
9. NIST, NIST: Smart Phone Tool Assessment Test Plan, National Institute of Standards and Technology (August 2009), http://www.cftt.nist.gov/mobiledevices.htm
10. Curran, K., Robinson, A., Peacocke, S., Cassidy, S.: Mobile phone forensic analysis. International Journal of Digital Crime and Forensics 2(2), 15–27 (2010)
11. Somasheker, A., Keesara, H., Luo, X.: Efficient forensic tools for handheld devices: A comprehensive perspective. In: Proceedings of Southwest Decision Sciences Institute (March 2008)
12. Kim, K., Hong, D., Chung, K., Ryou, J.-C.: Data acquisition from cell phone using logical approach. World Academy of Science, Engineering and Technology 32 (2007)
13. Thing, V., Ng, K.-Y., Chang, E.-C.: Live memory forensics of mobile phones. In: Proceedings of DFRWS (2010)
14. Mokhonoana, P.M., Olivier, M.S.: Acquisition of a symbian smart phone's content with an on-phone forensic tool. In: Proceedings of the Southern African Telecommunication Networks and Applications Conference (SATNAC) (September 2007)
15. Connor, T.P.O.: Provider side cell phone forensics. Small Scale Digital Device Forensics Journal 3(1) (2009)
16. Distefano, A., Me, G., Pace, F.: Android anti-forensics through a local paradigm. Digital Investigation, 95–103 (2010)
17. Garfinkel, S.: Anti-forensics: Techniques, detection and countermeasures. In: Proceedings of the 2nd International Conference on i-Warfare and Security (ICIW), Monterey, CA (March 2007)
18. Test Results for Mobile Device Acquisition Tool: Secure View 2.1.0 (November 2010), http://ncjrs.gov/pdffiles1/nij/232225.pdf
19. Test Results for Mobile Device Acquisition Tool: XRY 5.0.2 (November 2010), http://ncjrs.gov/pdffiles1/nij/232229.pdf
20. Test Results for Mobile Device Acquisition Tool: Device Seizure 4.0 (November 2010), http://ncjrs.gov/pdffiles1/nij/232230.pdf
21. Test Results for Mobile Device Acquisition Tool: CelleBrite UFED 1.1.3.3 - Report Manager 1.6.5 (November 2010), http://ncjrs.gov/pdffiles1/nij/231987.pdf

22. Cellebrite mobile data secured, http://www.cellebrite.com/
23. Corporation, P.: http://www.paraben.com/
24. Micro Systemation XRY application, http://www.msab.com/xry/current-version-release-information
25. Guidance Software EnCase Neutrino, http://www.encase.com/products/neutrino.aspx
26. Susteen SecureView, http://www.secureview.us/secureview3
27. Harris, R.: Arriving at an anti-forensics consensus: Examining how to define and control the anti-forensics problem. In: Proceedings of Digital Forensic Research Workshop (2006)
28. Professional Software USB Protocol Analyzer, http://www.usblyzer.com/

Mobile Real-Time Tracking System Based on the XCREAM (XLogic Collaborative RFID/USN-Enabled Adaptive Middleware)

Je-Kuk Yun, Kyungeun Park, Changhyun Byun, Yanggon Kim, and Juno Chang

Abstract. The main objective of this paper is to provide people who use smartphones with a convenient way of getting useful information by making their smartphones communicate with the XCREAM (XLogic Collaborative RFID/USN-Enabled Adaptive Middleware). We have developed the XCREAM, which is a smart mediator between many types of sensor devices and applications. It gathers a variety of raw data from many types of USN devices, such as temperature or humidity sensors and RFID readers. It also distributes the collected event data to the appropriate applications according to the pre-registered scenarios. In addition, we have been extending the external interface of the XCREAM, where the XCREAM receives data from smartphones as well as ordinary USN devices. Furthermore, the extension enables the XCREAM to deliver the sensor data to the corresponding smartphone(s) in order to make applications for the smartphone(s) be more flexible and interactive. In this paper, we present a simple but feasible scenario that helps a package receiver simply track the current location of its delivery truck for the package.

Keywords: XCREAM (XLogic Collaborative RFID/USN-Enabled Adaptive Middleware), RFID (Radio Frequency Identification), USN (Ubiquitous Sensor Networks); tracking; GPS (Global Positioning System).

1 Introduction

In the past decade, wireless network communication technology and many types of smartphones have been making great progress, quantitatively as well as

Je-Kuk Yun · Kyungeun Park · Changhyun Byun · Yanggon Kim
Towson University 8000 York Rd Towson, MD, 21093 U.S.A.
e-mail: {jyun4,kpark3,cbyun1,ykim}@towson.edu

Juno Chang
Sangmyung University 7 Hongji, Jongno, Seoul, 110-743 Rep. of Korea
e-mail: jchang@smu.ac.kr

R. Lee (Ed.): Software Eng. Research, Management & Appl. 2011, SCI 377, pp. 157–171.
springerlink.com

qualitatively. As smartphones have been pervaded, people are able to access the Internet and get useful information, anytime, anywhere by wireless. Besides, various specialized industries, such as manufacturing, logistics, and supply chains, also have started to introduce wireless network communication and RFID technologies to their business sites in order to get highly effective quality control and efficient logistics. RFID technology supports contact-less and wireless information access through radio waves for identifying, capturing, and transmitting information from tagged objects to enterprise-level systems. RFID technology was initially introduced to allow any applications based on tracking and tracing technology to capture identification information without the need for human intervention. These kinds of services need to be integrated into comprehensive services, as people expect more convenient and efficient services based on the existing technologies and services.

In order to provide these complex services, we have been developing the XCREAM [1, 2]. The XCREAM is able to reliably collect a huge number of sensor data and quickly respond to requests from applications according to the pre-defined scenario of various parties. The scenarios simply can be registered by users through web interface in the form of XLogic script and processed by the WAS (Web Application Service) Agent [3].

As remarkably advanced smartphone has become popular, we make a decision to allow the XCREAM to include it as a special case of self-activating sensor within the framework. Ultimately, we want various kinds of sensors and smartphones to issue sensing data and to trigger specific app applications running in the smartphones as well as the existing business application services.

To show the practical availability of this model, we developed mobile real-time tracking system based on XCREAM framework. The XCREAM periodically collects each delivery truck's current location information from the driver's smartphone, matches the package to be delivered, and notifies each individual recipient of the current location of the package. This system can help the receiver to estimate its time of arrival of the package and be ready to treat it properly.

In fact, it is hard to take full advantage of this kind of application, unless we can immediately get all the tracking information on the fly. Generally, most of the delivery companies provide recipients with tracking information through their websites online. In this environment, the receivers, however, can acquire limited or delayed tracking information of their packages and it makes them need to wait at home until its arrival, paying attention to the door or the doorbell. It may happen that the receiver misses the package when the delivery truck arrives ahead of schedule or behind the schedule.

This paper consists of six sections. In Section 2, we discuss related research required to build the mobile real-time tracking system. Then, in Section 3, we introduce the XCREAM and its components. Section 4 explains the mobile real-time tracking system and its detailed processes. Section 5 and Section 6 contain conclusion and future research, individually.

2 Related Work

This section summarizes the underlying technologies which are adopted in developing the XCREAM framework.

2.1 SOA

A service-oriented architecture has been chosen as a basic interface scheme for the application framework for multiple services. These services need to communicate with each other. The communication may involve data exchange among two or more services which are related with each other. The interface of the SOA is independent of the programming language, such as JAVA and C++, hardware platform, and the operating system, in which the service is implemented.

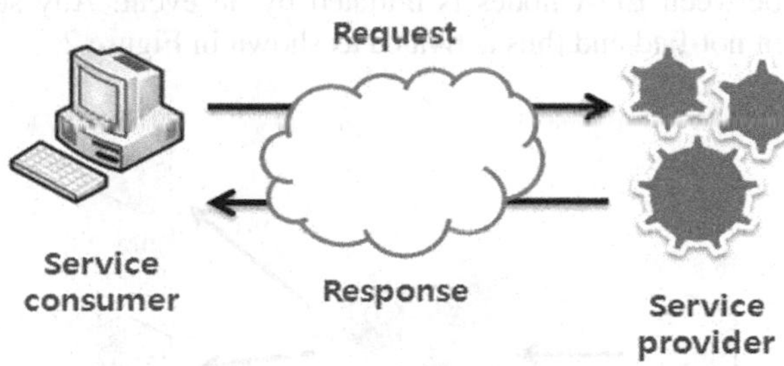

Fig. 1 Request/reply mechanism in a SOA

Figure 1 presents that the SOA is based on a conventional request/reply mechanism. A service consumer requests a service provider through the network and cannot receive the response until the completion of the operation on the provider side.

The main effect of the SOA scheme is a loose coupling of services with operating systems, programming languages and other application specific technologies [4]. SOA makes functions separated into distinct units, or services [5], which are accessible over a network in order that they can be combined and reused in the production of business applications [6]. Services communicate with each other by passing data from one service to another, or by coordinating an activity between two or more services. SOA concept is often seen as built upon, and evolving from older concepts of distributed computing and modular programming.

The XCREAM adopts the service-oriented architecture (SOA) as the interface scheme required to integrate RFID/USN-based application services not only to guarantee independence of individual services, but also to gain flexibility when extending the platform infrastructure by enabling new collaborative services to work with the platform or other application services. Basically, the SOA works through the following steps. Existing application functions are classified into distinct functional business units. They are reassembled into proper software units, namely, services, through standardized calling interfaces; and the services are combined in order to construct real-world business applications. Based on the SOA structure, the platform exposes customized web services as the method of establishing connections with heterogeneous external systems.

2.2 EDA

Gartner introduced a new methodology in 2003 to describe a design paradigm based on events, which is the Event Driven Architecture (EDA). EDA is a method to design and develop applications or systems where events transmit between separated software components and services. EDA complements the SOA. SOA is normally a better method for a request/response exchange but EDA introduces long-running asynchronous process capabilities. Furthermore, an EDA node transmits events and does not depend on the availability of service consumers. The communication between EDA nodes is initiated by an event. Any subscribers to that event are then notified and thus activated as shown in Figure 2.

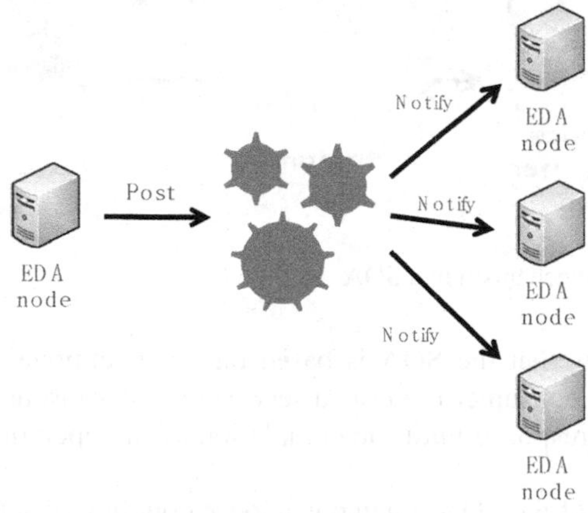

Fig. 2 Publish/subscribe mechanism in an EDA

The XCREAM framework is based on the EDA, which is completely decoupled from the specific RFID/USN middleware. This scheme enables the framework to work independently with an RFID/USN middleware and execute service scenarios, only when the relevant events from the RFID/USN middleware are recognized.

2.3 GPS

The Global Positioning System (GPS) is a navigation system based on satellite technology. And the GPS is developed by the US Department of Defense. The main function of the GPS is to locate a GPS reader based on position, velocity, and time determination system. And the GPS works 24 hours a day for free and offers high accurate location information. The GPS can guarantee sub-decimetre horizontal position accuracy (<10m), as long as signals from the constellation of GPS are not obstructed [7]. And the GPS technology consists of three discrete, as it is shown in the following Figure 3. They are satellites in orbit, ground control stations, and users (satellite receivers found in land, air, and sea).

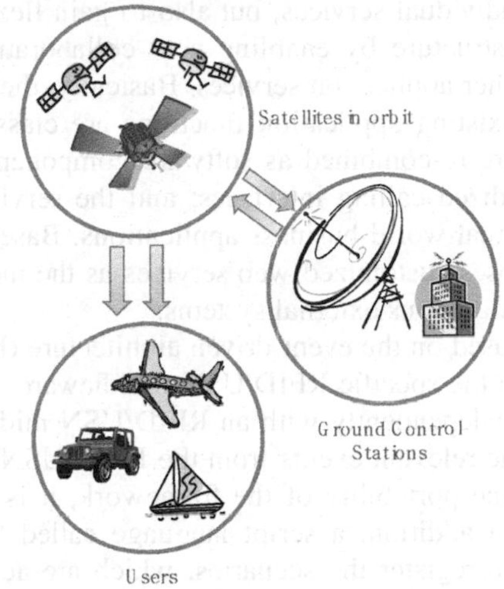

Fig. 3 The constituents of the GPS

The GPS equipment, especially a GPS reader, is combined into electronic devices such as a navigation device and even telecommunication chipset. It follows the introduction of various kinds of services like a tracking service based on positioning an object including person, vehicle and valuable assets [8, 9, 10]

In the mobile real-time tracking system, the driver's smartphone periodically captures the GPS information and send the information into the XCREAM. In the next section, the XCREAM will be introduced for further understanding of the whole framework.

3 The XCREAM Framework

3.1 The XCREAM

The proposed XCREAM middleware platform provides a scenario-based collaborative framework, which seamlessly integrates various application services in the RFID/USN environment. The XCREAM is composed of four agents such as the Collector Agent, the Proxy Agent, the Event Activation Agent, the Web Application Service Agent, and the Event Handler. Upper-level application services are to collaborate with each other, through the Event Handler, which allows pre-registered scripts to be executed depending on the event data collected by the XCREAM. The scripts are written in XML infrastructure language, called the XLogic, which describes specific service scenarios and interacts with the XCREAM framework. Each agent may exchange event data, under the supervision of the Event Handler.

The XCREAM adopts the service-oriented architecture (SOA) as an interfacing scheme for integrating RFID/USN-based application services not only to guarantee independence of individual services, but also to gain flexibility when extending the platform infrastructure by enabling new collaborative services to work with the platform or other application services. Basically, the SOA works through the following steps. Existing application functions are classified into functional business units. They are re-combined as software component units, namely, services, through standardized calling interfaces; and the services are combined in order to construct the real-world business applications. Based on the SOA structure, the platform exposes customized web services as the method of establishing connections with heterogeneous external systems.

The framework is based on the event-driven architecture (EDA), which is completely decoupled from the specific RFID/USN middleware. This scheme enables the platform to work independently with an RFID/USN middleware and execute scenarios only when the relevant events from the RFID/USN middleware are recognized. To promote the portability of the framework, it is written in Java programming language. In addition, a script language called "XLogic (eXtensible Logic)" allows users to register the scenarios, which are activated by a specific event and deliver the event to the related service applications. In order to support real-time processing, a caching system is also considered, which may keep XLogic scripts within memory, depending on their frequency of use. This scheme is expected to shorten the time for parsing the scripts.

3.2 The XCREAM Component

3.2.1 The Event Handler

The Event Handler is in charge of The Event Handler is in charge of managing the remaining four agents and delivering every event received from each agent to the appropriate parties. Internal events just pass through the Event Handler. The Event Handler propagates events to all the agents and then the agents accept only events of interest for further processing and ignore the remaining events.

3.2.2 The Collector Agent

Events originated from RFID/USN middlewares are collected by asking the XCREAM for the following queries: a snapshot query with which a RFID/USN middleware requests immediate real-time identification or sensor data and a continuous query that is kept in an active state and used to continuously request the data during a specified period.

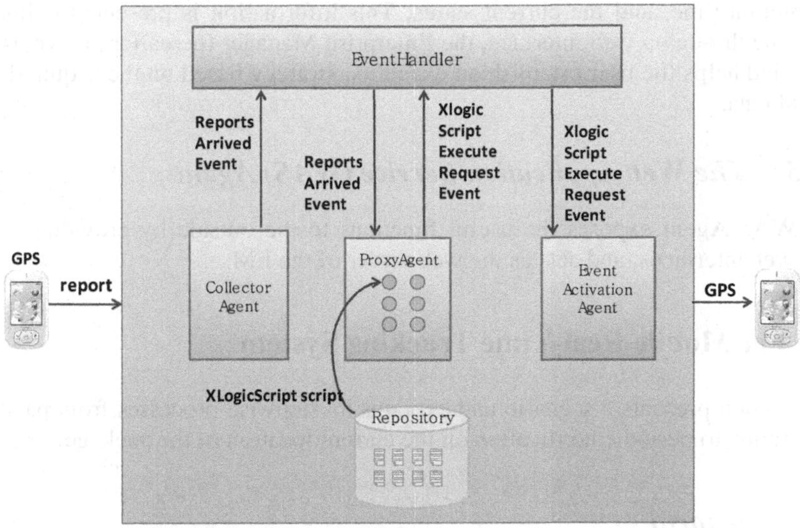

Fig. 4 Overall Event Flow of the XCREAM

The Collector Agent is connected to various RFID/USN middlewares through multi-threaded socket connections on the pre-allocated port and is in charge of forwarding the identification or sensor data to the Event Handler after converting them to the corresponding Java objects, called "ReportsArrivedEvent" (see Figure 4).

3.2.3 The Proxy Agent

The Proxy Agent is responsible for shortening the overall-event processing time. An XLogic script, which has been executed, is in memory as an executable Java object rather than the XML script itself saved in the repository. This buffering scheme remarkably reduces parsing time compared to the fetch-and-execution scheme from the repository and makes it possible to effectively manage the system resources throughout the whole lifetime of the XLogic script related with a specific service scenario.

The Proxy Agent usually extracts the unique identification information of "ReportsArrivedEvent," which is forwarded by the Event Handler. If the value equals to one of the XLogic script resident in memory, then the XLogic is converted into "XLogicScriptExecuteRequestEvent" and sent back to the Event Handler. If it is not found in memory, then the XLogic in XML form is queried and then parsed to be placed in memory as an executable Java object. Accordingly, the event is delivered to the Event Handler.

3.2.4 The Event Activation Agent

The Event Activation Agent not only executes the XLogic script but also keeps the statistics of the active XLogic such as the success and failure ratio, the last completion time, and the current status. This information is presented online in real-time through a web interface, the Enterprise Manager (hereafter known as the EM), and helps the user establish an execution strategy based on the acquired statistical data.

3.2.5 The Web Application Service (WAS) Agent

The WAS Agent exposes the useful functions to the outside by providing users with web interfaces, and acts as the web server of the EM.

4 The Mobile Real-Time Tracking System

This section presents a scenario and explains its stepwise processes from package registration to periodic notification on the current location of the package.

4.1 Scenario

The following figure illustrates a scenario in which a sender requests the FIRM X to deliver his/her packages.

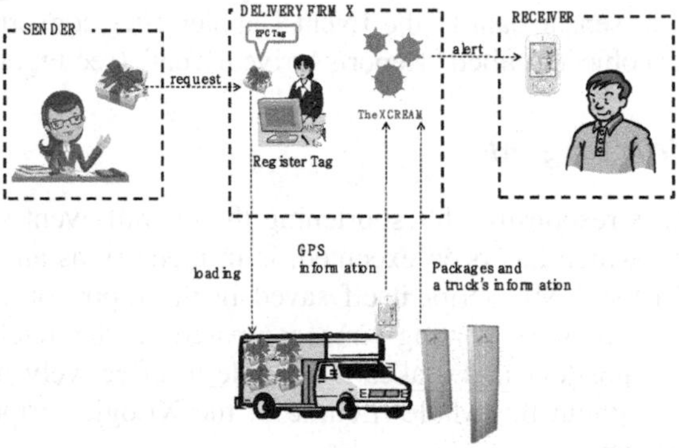

Fig. 5 Auto-tracking system

The FIRM X receives the packages with the recipient's information and the FIRM X receptionist puts an RFID tag on the package and registers it to the XCREAM. The packages with tags are loaded on to a delivery truck and when the truck passes the gate, the RFID readers scan the packages in the truck and send the captured tag information to the XCREAM, so that the XCREAM can relate the packages with the truck. As the truck is moving towards recipient, the tracking system helps a receiver be able to check its current location through his/her smartphone. As a result, the customer can estimate its expected time of arrival more precisely.

4.2 *Package Registration*

When a new package is dropped off at the FIRM X, a receptionist registers the package associated information, such as tracking number, recipient's information including his/her address and phone number, into the system. Figure 6 is the package registration interface of the system.

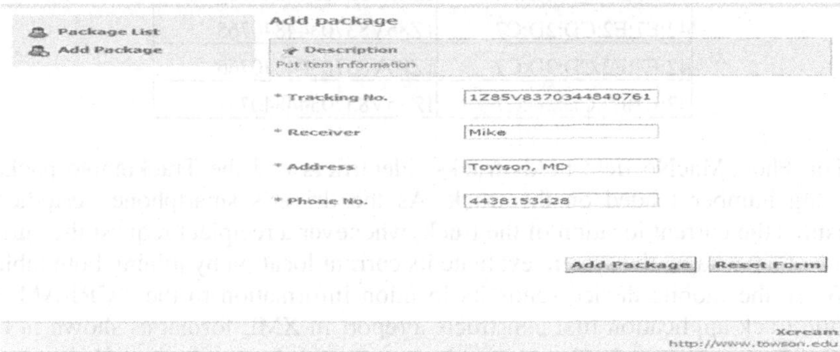

Fig. 6 Package registration

4.3 *Sending the GPS to the XCREAM*

The system keeps track of the packages on the basis of the location information of the truck driver. To do so, the system should know which packages are contained in each truck. The package information loaded on the truck is maintained as follows:

Table 1 Package Information Table

TrackingNo	Receiver	Address	Phone
1Z85V8370344840761	Mike	Towson, MD	4438153428
1Z85V8370344840762	James	Cockeysville, MD	4431545452
1Z85V8370344840763	Steve	Pikesville, MD	4431518484
1Z85V8370344840764	Julis	Catonsville, MD	4431548154
1Z85V8370344840765	Edi	Towson, MD	4431587651
1Z85V8370344840766	Angelina	Parkville, MD	4438789545
1Z85V8370344840767	Feti	Towson, MD	4438978453

When a truck passes through the gate, the RFID reader sends tag information to the XCREAM. Then the system automatically creates a new table as follow:

Table 2 Load Table

PhoneMacNo	TrackingNo
38:E7:D8:BA:8D:D7	1Z85V8370344840761
38:E7:D8:BA:8D:D7	1Z85V8370344840762
38:E7:D8:BA:8D:D7	1Z85V8370344840763
38:E7:D8:BA:8D:D7	1Z85V8370344840764
42:E7:E2:CD:2D:C7	1Z85V8370344840765
42:E7:E2:CD:2D:C7	1Z85V8370344840766
42:E7:E2:CD:2D:C7	1Z85V8370344840767

The PhoneMacNo field holds trucks' identifiers and the TrackingNo package tracking number loaded on the truck. As the driver's smartphone periodically transmits the current location of the truck, whenever a recipient request the current status of a package, the system evaluate its current location by joining both tables.

When the mobile device sends its location information to the XCREAM, the mobile truck application first constructs a report in XML format as shown in Figure 7. The report contains location information with longitude, latitude, date, and time.

```
<report name="report1" source="truck">
        <mac>38:E7:D8:BA:8D:D7</mac>
        <ip>10.55.17.48</ip>
        <date>04-25-2011</date>
        <time>15:27:36</time>
        <data>
                <name>longitude</name>
                <value>76.605713</value>
        </data>
        <data>
                <name>latitude</name>
                <value>39.390604</value>
        </data>
</report>
```

Fig. 7 Report of Location Information

4.4 The Collector Agent

The above figure shows the class diagram of the Collector Agent. The CollectorAgentImpl class processes three main tasks: collecting reports in XML format, converting the reports to the ReportsArrivedEvent, and analyzing reports. The CollectorClient is to offer multi-threaded socket connections to the various RFID/USN middlewares or smartphones so that they can send reports to the XCREAM.

The CollectorAgentImpl is to start or stop the CollectorClient and check the status of the Collector Agent that has the following four states: STARTED, STARTING, STOPPED, and STOPPING. It also converts the reports into the ReportsArrivedEvent so that other agents can process the ReportsArrivedEvent. Lastly, the CollectorStatistic distinguish the meaningful reports from lots of anonymous reports. It also counts incoming reports and analyzes the reports. After then, the CollectorAgent delivers the ReportsArrivedEvent to the Event Handler.

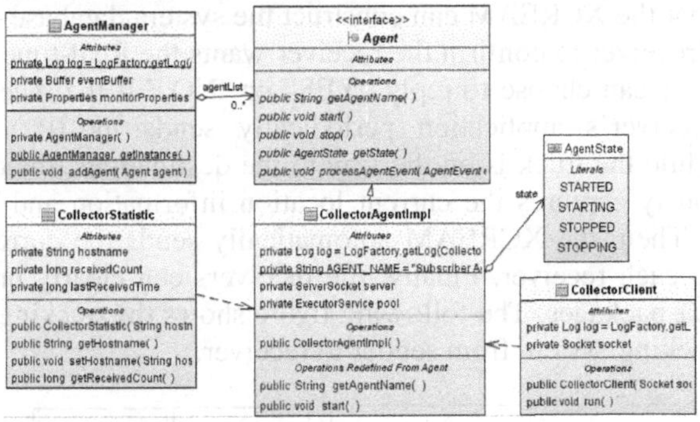

Fig. 8 Class Diagram of the Collector Agent

4.5 *TruckApp Application*

Figure 9 describes class diagram of the mobile truck application. The application takes part in sending the current location to the XCREAM. First, when the truck starts to deliver packages, the truck driver runs the TruckApp on the driver's smartphone. While the driver is on the way to each destination, the mobile application captures the location information in the form of the longitude and latitude. Upon capturing the current location, the MakingReport class builds a report in XML format. Finally, the SendingReport class connects communication path to the XCREAM and sends the report to the XCREAM.

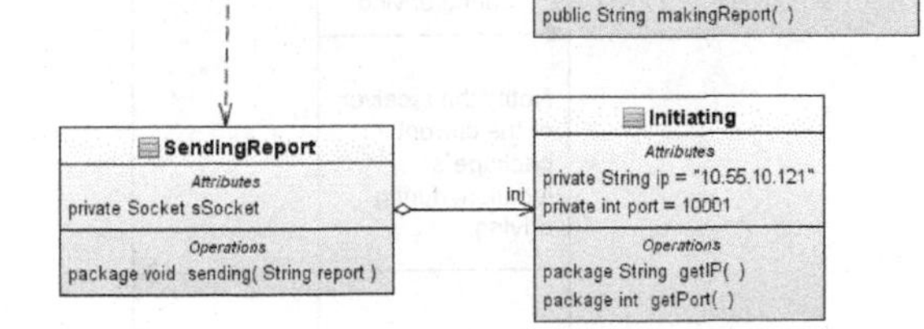

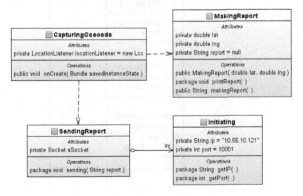

Fig. 9 Class Diagram of the Truck Application

4.6 Tracking Procedures

The sender drops a package to the FIRM X with the sender's information and the receiver's name, phone number, and address. The FIRM X puts an RFID tag on the package. A tracking number of the tag will be used to relate the package with the receiver's address, and the receiver's phone number. The tagged packages are loaded on to a delivery truck. When the truck passes through the main gate, the RFID reader scans the items in the truck and sends the tag information to the XCREAM so that the XCREAM can construct the system database. It also sends a message to the receiver to confirm the receiver wants the Real-time Tracking Service. The receiver can choose to reply "YES" or "NO." If the receiver wants the service, the receiver's application periodically sends its IP address to the XCREAM. While the truck is on the way to the destinations, the driver's smartphone continuously captures the current location information and sends them to the XCREAM. Then, the XCREAM automatically sends the current location of the packages to each receiver. Finally, the receivers can clearly predict the time of arrival of their packages. The following figure shows the tracking procedures of the real-time tracking system from sender to receiver.

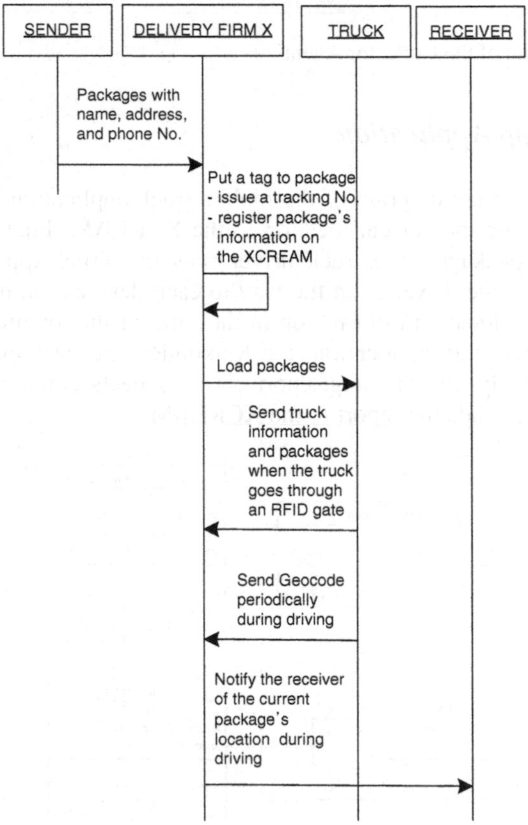

Fig. 10 Procedure from the sender to the receiver

4.7 Receiver Application

Receiver's application is implemented with java and uses the Google API. When the application receives GPS information, the receiver can check the current location of the delivery truck. The following figure demonstrates package tracking situations: the upper one denotes that the receiver is checking the current location of

Fig. 11 Simulation of the Receiver's Application

the package with receiver's smartphone; and the bottom one shows the current location of the delivery truck 5 minutes after than the previous status.

That application tells the receiver exact arrival time of the package because it works based on the GPS. Therefore, it can make the receiver to clearly predict the arrival time so the receiver will not miss it.

5 Conclusion

As wireless network communication technology and several types of smartphones have made great progress, many people have started to use a smartphone in order to not only take advantage of various useful applications but to also have access to useful information anytime, anywhere. Furthermore, the cost of the smartphone keeps going down and it is possible to download useful information at a low cost because wireless Internet access points have been diffused almost everywhere.

Many various industries have started to introduce wireless network communication technology for effective quality management and efficient logistics, as well as for supporting product safety. Nowadays, those systems are intricately related to each service to efficiently process multiple requests. In order to provide those complex services, we have been developing the XCREAM. The XCREAM is able to support the reliable collection of a large amount of sensor data and quick response to requests into applications, or even smartphones, according to the predefined scenario of individual services.

Now users can easily define scenario into the XCREAM with their smartphone and can get the real-time response from the XCREAM anytime, anywhere. We can clearly predict when our packages will arrive by checking our smartphone if we define a scenario to the XCREAM.

6 Future Work

There are a variety of raw data that the XCREAM can collect. For now, the XCREAM collects the data and sends them to the appropriate application. After that, the XCREAM just throws away the data. However, if the XCREAM stores those data into repository, the XCREAM can make a decision when data in the repository meet certain conditions. For example, the XCREAM counts the number of the customers by using an RFID reader in a restaurant where 50 people is of full capacity. When the restaurant is full, the XCREAM can issue an event to notify other customers of no vacant seats. In addition, the XCREAM will be able to manage inventory. If any items are missing or sold out, the XCREAM can automatically order that item or notify inventory manager of it. As I stated before, the XCREAM can provide further various benefits with users if the XCREAM can recognize a specific context and then make a decision.

References

1. Park, K., Kim, Y., Chang, J., Rhee, D., Lee, J.: The Prototype of the Massive Events Streams Service Architecture and its Application. In: Proceedings of the Intl. Conf. on Software Engineering, Artificial Intelligence, Networking and Parallel/Distributed Computing (SNPD 2008), Thailand (2008)
2. Park, K., Yun, J., Kim, Y., Chang, J.: Design and Implementation of Scenario-Based Collaborative Framework: XCREAM. In: Proceedings of the Intl. Conf. on Information cience and Application (ICISA 2010), Seoul, Korea (2010)
3. Park, K., Kim, Y., Lee, J., Chang, J.: Integrated Design of Event Stream Service System Architecture (ESSSA). In: Proceedings of the Intl. Conf. on E-business, Barcelona, Spain (2007)
4. Salz, R.: WSDL 2: Just Say No, O'Rielly xml.com (2004), `http://webservices.xml.com/pub/a/ws/2004/11/17/salz.html`
5. Little, M.: Does WSDL 2.0 Matter, InfoQ (2007), `http://www.infoq.com/news/2007/01/wsdl-2-importance`
6. W3C, Web Services Description Language (WSDL) 1.1, W3C Note (2001), `http://www.w3.org/TR/wsdl`
7. Ludden, B.: Location technology (2000), `http://www.locationforum.org/about-LIF/documents`
8. Byman, P., Koskelo, I.: Mapping Finnish roads with differential GPS and dead reckoning, GPS World, pp. 38–42 (February 1991)
9. Papaioannou, P., Tziavos, I.N.: The use of the global positioning system (GPS) for the collection of spatial information and the creation of data bases concerning the National Highway Network, Research project, Final Report, Thessaloniki (1993)
10. Tokmakidis, K., Tziavos, I.N.: Mapping of Edessa prefecture road network using GPS/GIS, Aristotle University of Thessaloniki, Internal Report (2000)

A Quantitative Evaluation of the Impact of Architectural Patterns on Quality Requirements

Mohamad Kassab[1], Ghizlane El-Boussaidi[1], and Hafedh Mili[2]

Abstract. When designing software architectures, an architect relies on a set of pre-defined styles commonly named architectural patterns. While architectural patterns embody high level design decisions, an architectural tactic is a design strategy that addresses a particular quality attribute. Tactics; in fact, serve as the meeting point between the quality attributes and the software architecture. To guide the architect in selecting the most appropriate architectural patterns and tactics, the interactions between quality attributes, tactics and patterns should be analyzed and quantified and the results should be considered as decision criteria within a quality-driven architectural design process. In this paper, we propose an approach for a quantitative evaluation of the support provided by a pattern for a given targeted set of quality attributes.

Keywords: architectural patterns, architectural tactics, quality requirements, architectural design.

1 Introduction

In the process-oriented and goal-oriented software development approaches, user requirements are established based on an analysis of business goals and of the application domain. Subsequently, architectures of the desired systems are designed and implemented. One of the primary goals of the architecture of a system is to create a system design to meet the required functionality, while satisfying the desired quality. In fact, if we consider a system's architecture as a set of architectural decisions, the most significant ones concern the satisfaction of quality attributes [1].

When designing software architectures, an architect relies on a set of idiomatic patterns commonly named architectural styles or patterns. A Software Architectural Pattern defines a family of systems in terms of a pattern of structural organization and behavior [3]. More specifically, an architectural style determines the vocabulary of components and connectors that can be used in instances of that style, together with a set of constraints on how they can be combined [2]. Many

[1] École de Technologie Supérieure, Montreal, Canada.
[2] Université du Québec à Montréal, Montreal, Canada.

R. Lee (Ed.): Software Eng. Research, Management & Appl. 2011, SCI 377, pp. 173–184.
springerlink.com

common architectural patterns are described in [3, 4, 7, 10, 19, 20]. Common architecture patterns include Layers, Pipes and Filters, Model View Controller (MVC), Broker, Client-Server and Shared Repository.

While architectural patterns embody high level design decisions, an architectural tactic [13] is a design strategy that addresses a particular quality attribute. Indeed tactics serve as the meeting point between the quality attributes and the software architecture. In [8], the authors define an architectural tactic as an architectural transformation that affects the parameters of an underlying quality attribute model. "The structure and behavior of tactics is more local and low level than the architectural pattern, and therefore must fit into the larger structure and behavior of patterns applied to the same system" [9]. Implementing a tactic into a pattern may affect the pattern by modifying some of its components, adding some components and connectors, or replicating components and connectors [9].

Architectural patterns have been described using different frameworks (see e.g. [10], [11], [20]). A common framework to describing patterns includes the name, an example, the problem, the structure and the dynamics of the solution, and the consequences of using a pattern expressed in terms of its benefits and its liabilities. Hence the selection of an architectural pattern is qualitatively driven by the properties it exhibits [17] and the architecture knowledge and experience with patterns. To guide the architect in selecting the most appropriate architectural patterns and tactics, the interactions between quality attributes, tactics and patterns should be analyzed and quantified and the results should be considered as decision criteria within a quality-driven architectural design process. In this paper, we propose an approach for a quantitative evaluation of the support provided by a pattern for a given targeted set of quality attributes.

The paper is organized as follows. We start by the quantitative mapping of quality attributes into architectural tactics where we consider not only the positive contribution of tactics to implement the qualities but also the potential conflict among qualities due to the introduced tactics (Section 2), then we relate the architectural patterns to the tactics based on quantitative evaluations of the impact of incorporating the latter within the former (Section 3). The quantitative impact of incorporating the quality into architectural patterns is then calculated through a matrix transformation method using the data from the quality-tactic and tactic-pattern quantified relations (Section 4). Section 5 discusses related work, and Section 6 provides conclusion and the future work discussion. The approach is demonstrated throughout the paper through the analysis of the impact of incorporating both "Performance" and "Security" qualities into Pipes-Filters, Layers, MVC and Broker patterns.

2 Quality-Tactics Interaction

Quality is "the totality of characteristics of an entity that bear on its ability to satisfy stated and implied needs" [12]. Software Quality is an essential and distinguishing attribute of the final product. Many approaches [12, 14, 22] classify software quality in a structured set of characteristics which are further decomposed into sub-characteristics. In [24], quality taxonomy was built out of many inputted approaches starting from the ISO 9126-1 [12] to define the root nodes for the quality taxonomy (External Quality, Internal Quality and Quality in Use).

Tactics on the other hand are measures taken to improve the quality attributes [13]. For example, introducing concurrency for a better resource management is a tactic to improve system's performance. Similarly, Authentication and Authorization are popular tactics to resist unwanted attacks on the system and improve the overall security. In [13], the authors list the common tactics for the qualities: Availability, Modifiability, Performance, Security, Testability and Usability.

Typically, systems have multiple important quality attributes, and decisions made to satisfy a particular quality may help or hinder the achievement of another quality attribute. The best-known cases of conflicts occur when the choice of a tactic to implement certain quality attribute contributes negatively towards the achievement of another quality. For example, decisions to maximize the system reusability and maintainability through the usage of "abstracting common services" tactic may come at the cost of the "response time". "Authentication" is a classical tactic to achieve Security, but it may come on the cost of Performance and Usability; as in the case of requiring additional ID. Therefore, architects must deal with tradeoffs, priorities and interdependencies among qualities and tactics before selecting some tactics that may achieve one quality and hinder another. In this paper, we present these interdependencies empirically using three point scale adopted from [14]; where +1 indicates that if the tactic is implemented then a positive contribution (support) is given to the quality; -1 indicates that if the tactic is implemented then a negative contribution (hinder) is given to the quality; and 0 indicates that the tactic has no impact on the quality.

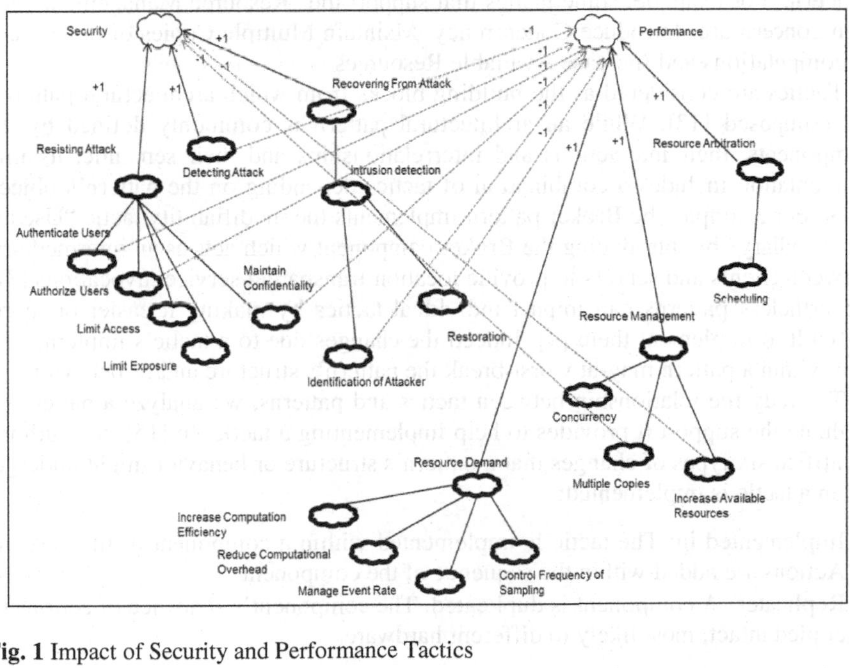

Fig. 1 Impact of Security and Performance Tactics

In Figure 1 we used the Softgoal Interdependency Graph notations [14] to illustrate the quality / tactics relationship on Security and Performance qualities. We used the same set of tactics recommended in [13]. In addition, we incorporated our observations towards the interdependencies between security / performance qualities due to the derived tactics.

For example, the Security tactic "Implementing Identification for Attacker" requires an audit trail, which is a space consuming; thus the value -1 was assigned between this tactic and Performance quality. The Security tactics "Authorization" and "Authentication" have negative effects on the response time of the system, and thus the value of -1 was assigned between these tactics and the Performance quality.

On the other hand, while "Maintaining Multiple Copies" tactic contributes positively to the achievement of performance, each copy increases the risk of revealing confidential data, and thus a value of -1 was assigned between this tactic and Security quality.

3 Tactics – Architectural Patterns Interaction

For a given quality attribute, the tactics are organized into sub-sets called categories in [13] and design concerns in [18]. Captured qualities can be realized through satisfying these design concerns. For example, performance is divided into three design concerns: resource demand, resource management and resource arbitration. A design concern by itself is realized through the implementation of a set of proper tactics. Each tactic is a design option for the architect [13] to satisfy a particular concern. For example, some tactics that support the "Resource Management" design concern are: Introduce Concurrency, Maintain Multiple Copies of either data or computations and Increase Available Resources.

Tactics are considered as the building blocks from which architectural patterns are composed [13]. While an architectural pattern is commonly defined by its components, their interactions and interrelationships and their semantic, its implementation includes a combination of tactics depending on the pattern's objectives. For example, the Broker pattern implements the modifiability tactic "Use an Intermediary" by introducing the Broker component which acts as an intermediary between clients and servers to provide location-transparent service invocations [8]. Nevertheless patterns can impact individual tactics by making it easier or more difficult to implement them [9]. Indeed the changes due to a tactic's implementation within a pattern may-at worst-break the pattern's structure and/or behavior.

To study the relationship between tactics and patterns, we analyze a pattern to evaluate the support it provides to help implementing a tactic. In [15], the authors identified six types of changes that a pattern's structure or behavior might undergo when a tactic is implemented:

- Implemented in: The tactic is implemented within a component of the pattern. Actions are added within the sequence of the component.
- Replicates: A component is duplicated. The component's sequence of actions is copied intact, most likely to different hardware.

- Add, in the pattern: A new instance of a component is added to the architecture while maintaining the integrity of the architecture pattern. The new component comes with its own behavior while following the constraints of the pattern.
- Add, out of the pattern: A new component is added to the architecture which does not follow the structure of the pattern. The added actions do not follow the pattern.
- Modify: A component's structure changes. This implies changes or additions within the action sequence of the component that are more significant than those found in "implemented in".
- Delete: A component is removed.

In this paper, we adopt the five point scale from [15] to identify the impact magnitude of incorporating tactics within architectural patterns, where +2 presents "Good Fit" category, +1 presents "Minor Changes" required for the incorporation, -1 presents "Significant changes" required for the incorporation, -2 presents "Poor Fit" category and 0 presents that the tactic and the pattern are basically orthogonal. The impact is defined as a function of number of participants impacted as listed in Table 1.

Table 1 Impact magnitude as a function of number of participants impacted [15].

Change Type	Number of Changes	Impact Range
Implemented in	1	+2 to +1
Replicates	3 or less	+2 to +1
	More than 3	+1 to 0
Add, in the pattern	3 or less	+2 to +1
	More than 3	+1 to 0
Add, out of the pattern	3 or less	0 to -1
	More than 3	-1 to -2
Modify	3 or less	+2 to -1
	More than 3	0 to -2

Tables 2 and 3 summarize our findings on the impact of implementing Security and Performance tactics (using the scale described above) on pipes and filters, Layered, MVC and Broker patterns.

For example, The Pipes and Filters architectural pattern provides a structure for systems that compute a stream of data [10]. Each computation is encapsulated in a component (filter). Data is passed between adjacent filters through connectors (pipes). Computations transform one or more input streams into one or more output streams, incrementally. There are no shared states between filters (stateless), and there is no central control in this pattern. In order to implement authorization, a new component must be introduced to manage role profiles along with new connections to each filter component. In addition, changes are required within existing filters to enable them to communicate with the new introduced component. These changes mean that significant work is required to implement authorization in this pattern; thus authorization is considered as a poor fit for the Pipes and Filters pattern as it represents a *major* "Add, out of the pattern" scenario and modifications to the existing components and thus -2 value was assigned.

Table 2 Impact of Incorporating Security Tactics into Pipes/Filters, Layered, MVC and Broker patterns

	Pipes / Filters	Layered Architecture	MVC	Broker
Security Tactics: Resisting Attacks				
Authenticate Users	-2	+1	-2	+2
Authorize Users	-2	-2	-2	-2
Limit Access	+1	+1	+1	-2
Limit Exposure	+2	+2	-2	-1
Maintain Data Confidentiality	-1	-1	-1	0
Security Tactics: Detecting Attacks				
intrusion detection	0	0	0	+1
Security Tactics: Recovering From an Attack				
Identification of Attacker	-2	-2	-2	-2
Restoration	-2	-1	0	+2
Average	-0.75	-0.25	-1	-0.25

Table 3 Impact of Incorporating Performance Tactics into Pipes/Filters, Layered, MVC and Broker patterns

	Pipes / Filters	Layered	MVC	Broker
Performance Tactics: Resource Demand				
increase computation efficiency	-1	-1	+1	+1
reduce computational overhead	-1	-1	+1	-2
manage event rate	-1	-1	+1	-2
control frequency of sampling	+1	+1	+1	+1
Performance Tactics: Resource Management				
introduce concurrency	+2	+1	+1	0
maintain multiple copies	-2	-1	+1	0
increase available resources	0	0	0	0
Performance Tactics: Resource Arbitration				
Scheduling Policy	-1	-1	+1	+1
Average	-0.375	-0.375	0.875	-0.125

Because of the "stateless" nature for the filters, this poses as a challenge when implementing the "restoration" tactic which aims at bringing the system to a latest correct state. In that case, it usually makes most sense to restart processing of that data from the beginning. If one must implement this tactic, then states must be defined for each filter and a mechanism must be created to restore a filter to the proper state when it comes back up. That may require a monitoring process. These changes involve major changes to components, plus possible *major* "Add, out of the pattern" and "Modify". Hence a value of -2 was assigned. On other hand, pipes-filters pattern naturally supports concurrent execution. Each filter can be implemented as a separate task and potentially executed in parallel with other filters [3]. The performance may be improved with filters running on different processors and processing data in an incremental and parallel way, i.e. a filter can work on partial results of its predecessors [16]. A value of +2 was assigned.

Regarding the Layers pattern, this is a widely used pattern. The Layers pattern structures an application by decomposing it into groups of subtasks where each group of subtasks is at a particular level of abstraction [10]. In other words, Layered Architecture is an organized hierarchy, each layer providing service to the layer above it and serves as a client to the layer below [3]. Because of Layers architecture organization, implementing the "Authenticate users" tactic is relatively easier than in Pipes and Filters. Indeed this tactic can be added as an additional layer on top of the existing ones; changes to the existing architecture are limited to its top layer and the new layer is within the structure of the pattern [15]. This is an example of (Add, in the pattern); and thus (+1) was assigned as a value. On other hand, as Layered Pattern supports layers of access, this limits the exposure by not placing all the logic and data into one layer. The spreading of layers among different hosts is recommended though to minimize the probability of an attack. A value of +2 was assigned for the difficulty of incorporating Limit Exposure in Layered architecture.

The MVC pattern divides an interactive application into three components [10]: 1) The model which encapsulates the core functionality and data; 2) Views which display the model information to the user; and 3) Controllers which handle user input and which are associated to views. The consistency between the model and its views is ensured using a change-propagation mechanism. Because the model component encapsulates core data and functionality, in order to implement the "Limit Exposure" tactic, this component needs to be broken into different components each with limited services. The components which will assume the model are preferably to be distributed among different hosts to limit the possibility of attacks. Such an implementation changes the pattern structure and behavior ("Major", Add, out of the pattern) and thus a value of -2 is assigned. On other hand, depending on the interface of the model, a view may need to make multiple calls to obtain all its display data [10]. This negatively impacts the response time to user requests and, hence to the Performance requirement. The response time may be improved by caching data within views. This is a "Minor" modification to an existing component; thus a value of +1 was assigned to "Manage Event Rate" tactic.

4 Calculation of the Impact of Architectural Patterns on Quality Attributes

We will now compute the quantitative impact of architectural patterns on quality attributes combining the data from the quality-tactic and tactic-pattern quantified relations. In the previous section, we discussed the latter impact considering each quality in isolation from the rest of the set of qualities that are aimed for the system. Averaging the values of impact of tactics against the patterns in the above analysis yields a value ranging from -2 to 2 representing how much the pattern is accommodating for the quality's tactics in abstract; with the value of -2 represents the least accommodating case, and the value of +2 represents the most accommodating.

For example, averaging the values for impact of Security tactics against the four patterns in Table 2 revealed that:

- Both Layered and Broker patterns are the most accommodating patterns for Security tactics.
- MVC is the least accommodating pattern for Security tactics.

Similarly for Performance, averaging the values for impact of Performance tactics against the four patterns in Table 3 revealed that:

- MVC is the most accommodating pattern for Performance tactics.
- Both Pipes/Filters and Layered are the least accommodating patterns for Performance tactics.

While this analysis is important for the architect as it provides quantitative evidences that can be used to optimize his architectural choices, it is still ignorant to the positive/negative interdependencies in the quality-tactics relationship as a major dimension to be considered when it comes to the architect's choices. We propose to consider the quality-tactics relation in conjunction with tactics-patterns relation through a matrix multiplication operation that yields the impact of architectural patterns on quality attributes.

Let A be the matrix of the impact values of the tactics on architectural patterns; A is composed of a set of a_{tp} values where a_{tp} is the impact value for the tactic t on the pattern p. In addition, let B be the matrix representing the quality-tactics relationship; B is composed of a set of b_{qt} values where b_{qt} represents the impact value of the tactic t on the quality q. Then the matrix representing the impact of incorporating quality attributes in architectural patterns is calculated as:

$$C = (AB) / \|\text{Total number of Tactics}\|$$

An element in C (e.g. c_{ij}) is the scalar product of the ith row of A with the kth column of B; divided by the total number of tactics being considered for the system.

The results from this calculation represent quantitatively the impact against each pattern p for each quality q (defined in total set of qualities Q being considered for the system) in the presence and relativity to the rest of all other qualities defined in set Q. It is important to note that the numbers in the result matrix (quality vs. pattern) are relative to the rest of the set of qualities to be considered for the system. That is, if we choose to consider adding more qualities to the system, for example, then these numbers will change accordingly.

Table 4 shows the results of our calculation following the above method on the impact of incorporating both Security and Performance Tactics into Pipes-Filters, Layered, MVC and Broker patterns. The results reveal that when both Security and Performance are to be considered as the key qualities of the system, then:

- MVC is the most accommodating pattern for the Performance quality while implementing all tactics of both Security and Performance in this system (it scored the highest among the four patterns against Performance).
- Broker is the least accommodating pattern for the Performance quality while implementing all tactics of both Security and Performance in the system (it scored the lowest among all patterns against Performance).
- Layered is the most accommodating pattern for the Security quality while implementing all tactics of both Security and Performance in this system (it scored the highest among the four patterns against Security).
- MVC is the least accommodating pattern for the Security quality while implementing all tactics of both Security and Performance in the system (it scored the lowest among all patterns against Security).

Table 4 Impact of Incorporating Security / Performance into selected architectural patterns.

	Security	Performance
Pipes-Filters	-0.25	0.3125
Layered	-0.063	0.125
MVC	-0.562	0.875
Broker	-0.125	-0.125

These results are based on the assumption that all the tactics implementing each quality are to be accommodated for the system. In practice; though, the architect may choose to select only a subset of the tactics to satisfy the quality achievement. The numbers derived from our approach can be useful for the architect to tune his selection of tactics as will be discussed in the future work section.

5 Related Work

Many patterns have been proposed and described in the literature (e.g. [3], [10], [4], [11], [20]). However, patterns consequences descriptions are incomplete, not searchable or cross-referenced and, mostly qualitative [9]. They are often classified according to the specific classes of systems they enable to construct (e.g. virtual machines). Our work aims at providing a quantitative method for comparing and evaluating architectural design choices and in particular by using information drawn from tactics descriptions as given in [13] and the quantitative impact of these tactics on quality attributes and architectural patterns. The SEI Attribute

Driven Design method [13] [18] describes an iterative process to designing software architecture where at each iteration of the process, the architect chooses architectural tactics and patterns that satisfy the most important quality attributes for that iteration. However the architectural decisions such as the selection of a pattern is mainly based on a qualitative evaluation and the architect expertise and knowledge.

Our analysis is based on the scale introduced in [15]. Harrison and Avgeriou propose a general model that relates patterns to tactics and quality attributes. In particular the proposed framework concentrates on the interaction between tactics and patterns and especially how tactic implementations affect patterns. Our approach may be seen as complementary to this framework as we consider both the impact of a tactic on a pattern and the impact of the tactic on other quality attributes.

Our work is closely related to Bode and Riebisch's work [17]. They present a quantitative evaluation of a set of selected architectural patterns regarding their support for the evolvability quality attribute. They refine the evolvability attribute into sub-characteristics and relate them to some properties that support good design. The selected patterns are used in a case study and the resulting design is assessed to determine the impact of these patterns on these properties. A matrix of impact values of patterns on evolvability sub-characteristics is inferred from the matrix relating properties to sub-characteristics and the matrix relating properties to patterns. For each pattern, the impact on evolvability is calculated from the resulting matrix as an average of all impact values on evolvability sub-characteristics. Our approach differs from Bode and Riebisch's approach in the way the analysis of the relationship between patterns and quality requirements was carried out; they use a particular case study and an evaluation by experts while our approach is based on the analysis of the generic structures and behavior of tactics and patterns. Besides, we consider more than one quality attribute in our analysis.

6 Conclusion and Future Work

In this paper, we proposed a quantitative approach to selecting architectural patterns starting from a subset of quality requirements. This approach relies on a quantitative assessment of the impact of architectural tactics on quality requirements, in the one hand, and the impact of incorporating these tactics in architectural patterns, in the other hand. We illustrate the approach using four common architectural patterns and assessing their support for both Security and Performance quality requirements. Though this is a preliminary quantitative investigation of the architectural patterns when considering more than one quality requirement, we believe that it's a key step towards a quality-driven architectural design process.

In the future, we plan to extend the ontology in [23] to integrate our quantitative approach to support the selection process of a set of tactics and patterns that satisfies a set of quality attributes while considering the trade-offs among quality attributes, tactics and patterns. Furthermore we plan to refine the approach in two ways. First we would like to refine our analysis and results by considering sub-characteristics of quality attributes (e.g. analyzing availability and confidentiality as sub-characteristics of security). Second, the numerical value that is assigned to a pattern regarding its support for a quality attribute depends on the selected subset of tactics to achieve the targeted attribute. In this paper we derived this value by considering all the tactics related to an attribute. However, in the future we would like to give the opportunity to an architect to tune these values by considering or discarding alternative tactics. This will help to alleviate impacts of the pattern whose choice was driven by a core of quality attributes on the other attributes.

References

[1] Jansen, J., van der Ven, J., Avgeriou, P., Hammer, D.K.: Tool Support for using Architectural Decisions. In: Proceedings of the 6th Working IEEE/IFIP Conference on Software Architecture (WICSA 2007), Mumbai, India (2007)

[2] Microsoft Application Architecture Guide: Patterns & Practices, 2nd edn., `http://msdn.microsoft.com/en-us/library/ff650706.aspx`

[3] Garlan, D., Shaw, M.: An Introduction to Software Architecture, Technical Report, CMU, Pittsburgh, PA, USA (1994)

[4] Avgeriou, P., Zdun, U.: Architectural Patterns Revisited– a Pattern Language. In: Proceedings of 10th European Conference on Pattern Languages of Programs (EuroPLoP 2005), pp. 1–39 (2005)

[5] Garcia, A.F., Rubira, C.M.F.: An architectural-based reflective approach to incorporating exception handling into dependable software. In: Romanovsky, A., Cheraghchi, H.S., Lee, S.H., Babu, C. S. (eds.) ECOOP-WS 2000. LNCS, vol. 2022, pp. 189–206. Springer, Heidelberg (2001)

[6] Garcia, A.F., Rubira, C.M.F., Romanovsky, A.B., Xu, J.A.: A Comparative Study of Exception Handling Mechanisms for Building Dependable Object-Oriented Software. Journal of Systems and Software 59(2), 197–222 (2001)

[7] Issarny, V., Benatre, J.-P.: Architecture-based Exception Handling. In: Proceedings of the 34th Annual Hawaii International Conference on System Sciences, Hicss-34 (2001)

[8] Bachmann, F., Bass, L., Nord, R.: Modifiability Tactics, Technical Report, SEI, CMU/SEI 2007-TR-002 (September 2007)

[9] Harrison, N., Avgeriou, P., Zdun, U.: On the Impact of Fault Tolerance Tactics on Architecture Patterns. In: Proceedings of 2nd International Workshop on Software Engineering for Resilient Systems (SERENE 2010), London, UK (2010)

[10] Buschmann, F., Meunier, R., Rohnert, H., Sommerlad, P., Stal, M.: Pattern-Oriented Software Architecture: A System of Patterns. John Wiley & Sons, Chichester (1996)

[11] Gamma, E., Helm, R., Johnson, R., Vlissides, J.M.: Design Patterns: Elements of Reusable Object-Oriented Software systems. Addison-Wesley, Reading (1994)

[12] International Standard ISO/IEC 9126-1, Software engineering – Product quality – Part 1: Quality model. ISO/IEC 9126-1:2001, 200 (2001)

[13] Bass, L., Clements, P., Kazman, R.: Software architecture in practice. Addison-Wesley, Reading (2003)

[14] Chung, L., Nixon, B.A., Yu, E., Mylopoulos, J.: Nonfunctional Requirements in Software Engineering. Kluwer Academic Publishing, Dordrecht (2000)

[15] Harrison, N.B., Avgeriou, P.: How do architecture patterns and tactics interact? A model and annotation. Journal of Systems and Software 83(10), 1735–1758 (2010)
[16] Gröne, B., Keller, F.: Conceptual architecture patterns. FMC-based representation. Univ.-Verl, Potsdam (2004)
[17] Bode, S., Riebisch, M.: Impact Evaluation for Quality-Oriented Architectural Decisions Regarding Evolvability. In: The 4th European Conference on Software Architecture, pp. 182–197 (2010)
[18] Wood, W.G.: A Practical Example of Applying Attribute-Driven Design (ADD), Version 2.0, Technical report CMU/SEI-2007-TR-005 ESC-TR-2007-005 (February 2007)
[19] Völter, M., Kircher, M., Zdun, U.: Remoting Patterns: Foundations of Enterprise. In: Internet and Realtime Distributed Object Middleware, Wiley, Chichester (2005)
[20] Shaw, M., Garlan, D.: Software Architecture: perspectives on an emerging discipline. Prentice Hall, Englewood Cliffs (1996)
[21] Yoder, J., Barcalow, J.: Architectural Patterns for Enabling Application Security. In: Proceedings of Pattern Language of Programs Conference, Allerton Park, Illinois, U.S.A (1997)
[22] Boehm, B.W., Brown, J.R., Lipow, M.: Quantitative Evaluation of Software Quality. In: The 2nd International Conference on Software Engineering, pp. 592–605. IEEE Computer Society, Los Alamitos (1967)
[23] Kassab, M.: Non-Functional Requirements: Modeling and Assessment. VDM Verlag Dr. Mueller (2009), ISBN 978-3-639-20617-3

Performance of Non-coherent Detectors for Ultra Wide Band Short Range Radar in Automobile Applications

Purushothaman Surendran, Jong-Hun Lee, and Seok Jun Ko

Abstract. A detector is said to be superior if the information is extracted in a best way for some particular purpose. Here the objective of this paper is to analyze the detection performance of non-coherent detectors such as square law detector, linear detector and logarithmic detector in a background of white Gaussian noise (WGN) for Ultra Wide Band Short Range Radar in Automotive applications. It is assumed that the target is stationary and energy reflected from the target is distributed. The non-coherent detectors have almost similar detection probability in a background of white Gaussian noise. The performance of the detector is analyzed and simulation has been done in order to verify.

Keywords: UWB Radar, Coherent Integration, Target Detection.

1 Introduction

The demand for short range radar sensors which are used for target detection has increased fabulously in automotive sector [2, 3]. The key ideas behind the development of Short Range Radar (SRR) in automotive sector are collision avoidance and reduce traffic fatality. The source for target detection is the radar signals reflected by the target, the received radar signal is a mixture of noise and varied signals. The designed system must provide the optimal technique to obtain the desired target detections, preferred detection can be determined by using specific algorithm for measuring the energy of the signals received in the receiver side. Decision is made on the basis of the received echo signal which is determined by target geometry. The detection algorithm is used to make a decision on the target present and to measure the range of the target. In order to predict the range more exactly, a larger signal bandwidth is required. This can be accomplished by reducing the range resolution

Purushothaman Surendran · Seok Jun Ko
Department of Electronics Engineering, Jeju National University, Jeju, Korea
e-mail: sjko@jejunu.ac.kr

Jong-Hun Lee
DGIST(Daegu Gyeongbuk Institute of Science & Technology), Daegu, Korea

R. Lee (Ed.): Software Eng. Research, Management & Appl. 2011, SCI 377, pp. 185–195.
springerlink.com

cell of the radar so that the range accuracy is increased. Consequently, the range resolution cell size of the radar is smaller than the target size and target is surrounded by homogeneous distributed noise [4].

Previous work [4, 5] has been mainly focused on the influence of increasing range resolution on the detection ability of targets with dimensions greater than the resolution cell. Also mathematical analysis of these circumstances has been resolved.

In this paper we analyze the performance of various detectors for Ultra Wide Band Short Range Radar (UWB-SRR) in automotive applications. It is assumed that the target is stationary and the energy reflected from the target is assumed to be 1. The performance of the detectors is shown.

The organization of this paper is as follows. In Section II, the system model is described. In section III, description about non-coherent detector. In section IV, the probability of detection and false alarm is expressed. In Section V, simulation results for various detectors. Finally, conclusions are presented in Section VI.

2 System Description

The block diagram of a UWB radar system as shown in fig. 1 is split into two parts, the transmitter part and the receiver part.

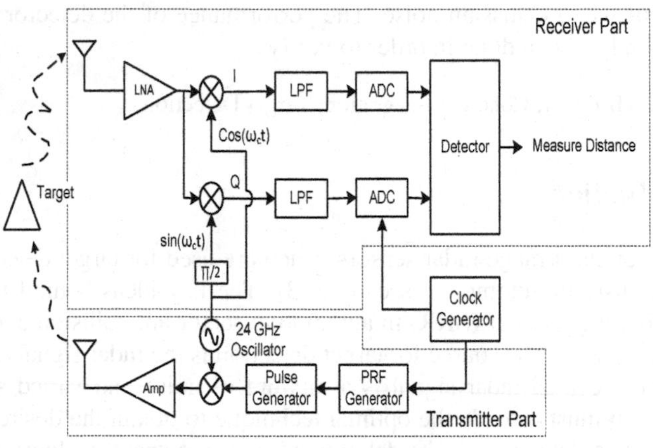

Fig. 1 Block Diagram of a UWB radar system

In the transmitter part, the pulses are initiated by the Pulse Repetition Frequency (PRF) generator which triggers the pulse generator which in turn generates Gaussian pulses with sub-nano second duration as shown in fig. 2. The Pulse Repetition Interval (PRI) is controlled by the maximum range of the radar. The maximum range for unambiguous range depends on the pulse repetition frequency and can be written as follows

$$R_{\max} = \frac{c}{2 \cdot f_{PRF}} \tag{1}$$

where f_{PRF} is pulse repetition frequency and c is the velocity of light. And the range resolution can be written as

$$\Delta R = \frac{c \cdot T_P}{2} \tag{2}$$

where T_P is pulse width and c is the velocity of light. And then the transmitted signal can be written as follows

$$s(t) = A_T \cdot \sin(2\pi f_c t + \varphi_0) \cdot \sum_{n=-\infty}^{+\infty} p(t - n \cdot T_{PRI}) \tag{3}$$

where $p(t)$ is the Gaussian pulse as follows

$$p(t) = \exp\left[-2\pi\left(\frac{t}{\tau_p}\right)^2\right] \tag{4}$$

where τ_p represents the time normalization factor, A_T is the amplitude of single transmit pulse, φ_0 is the phase of the transmit signal, f_c is the transmit frequency, T_{PRI} is the pulse repetition interval obtained from pulse repetition frequency given as $T_{PRI}=1/f_{PRF}$.

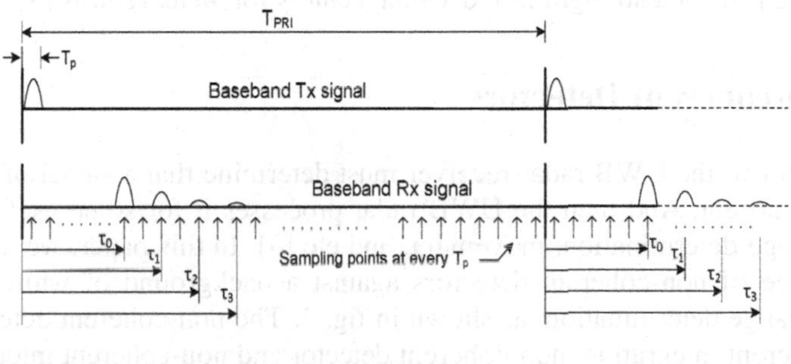

Fig. 2 Transmitted signal and received baseband signal

The form of the transmitted signal in this system is known, but the received signal usually is not completely known. Since the range resolution of this UWB radar system is much less than the extent of the target it must detect, the echo signal is the summation of the time-spaced echoes from the individual scattering centers that constitute the target [3]. In this paper, we assume that the target is stationary and the target has L independent reflecting cells. Then the target model is written as

$$h(t) = \sum_{l=0}^{L-1} \alpha_l \cdot \delta(t - \tau_l) \tag{5}$$

where the number of scatters L, the amplitude of the scatters α_l, and the time delays of the scatters τ_l are all unknown. We assume that the τ_0 in fig. 2 indicates the target range.

The radiated electromagnetic signals generated by the transmit antenna is reflected by the target and they are received in the receiver antenna. First, the received signal is pre -amplified by the Low Noise Amplifier (LNA). Then the signal is multiplied with carrier signal and divided between the in-phase and quadrature-phase baseband signal as shown in fig. 1. We assume that the low pass filter will match the pulse envelope spectral shape as close as possible to provide a fully matched optimum filter. Then the baseband received signal $r(t)$ is written as

$$r(t) = A_T \sum_{n=-\infty}^{+\infty} \sum_{l=0}^{L-1} \alpha_l \cdot e^{j\theta_l} p(t - nT_{PRI} - \tau_l) + n(t) \tag{6}$$

where $n(t)$ is the white Gaussian noise (WGN) with two-sided power spectral density $N_0/2$ and θ_l is the arbitrary phase of l-th scatter that can be written as $\theta_l = -2\pi f_c \tau_l + \varphi_0$. The sampling rate of the A/D converters is same to the pulse width. And we assume that the baseband received signal is sampled at peak point of $p(t)$ as like the fig. 2. When the target size is greater than the radar resolution, then the echo consists of a signal characterized by eq. (6) and fig. 2; the received echo signal will be composed of individual short duration signals in a pulse train. A gain rather than a loss can be obtained when a high-resolution waveform resolves the individual scatters of a distributed target such as the UWB radar system. Because there will be no signal addition from close scattering centers, detection will be depend on the reflected strength of individual centers for weak returns [3].

3 Non-coherent Detectors

The detector of the UWB radar receiver must determine that a signal of interest is present or absent. And then the UWB radar processes it for some useful purpose such as range determination, movement, and etc [3]. In this paper, we analyze the performance of non-coherent detectors against a background of white Gaussian noise for range determination, as shown in fig. 3. The non-coherent detectors consist of coherent integration, non-coherent detector and non-coherent integration.

The in-phase (I) and quadrature (Q) sampled values at every T_p are used as the input of the detector. It is assumed that the sampling rate (T_p) is same to the pulse width of 2 ns and the range resolution can be 30cm from (2). Also it is assumed that the maximum target range is 30m and by using (1) we can get T_{PRI} of 200 ns. From the above-mentioned range resolution and maximum target range, the range gates of at least 100 are required to detect the target. It is equal to the number of memory in the coherent and non-coherent integration. The sampled value at every T_p is applied to the switch I of the coherent integration. The switch I is shifted at every T_p sample, i.e., the range gate. It takes $N \cdot T_{PRI}$ to coherently integrate and dump for all of the range gates.

The coherent integration for the i-th range gate in I branch can be expressed as follows

$$X^I(i) = \frac{1}{N} \sum_{n=1}^{N} \mathrm{Re}\{r_n(iT_P)\} \tag{7}$$

where

$$r_n(iT_P)=A_T\sum_{l=0}^{L-1}\alpha_l\cdot e^{j\theta_l}p((nT_{PRI}+iT_p)-nT_{PRI}-\tau_l)+n'(nT_{PRI}+iT_P) \tag{8}$$

and where $n'(nT_{PRI}+iT_p)$ is a sampled value of $n(t)$ at $nT_{PRI}+iT_p$. Then the summations for each gate are stored in each memory of the coherent integration. Therefore it is possible to achieve an improvement of signal-to-noise ratio (SNR) as much as 10·log(*N*) [dB].

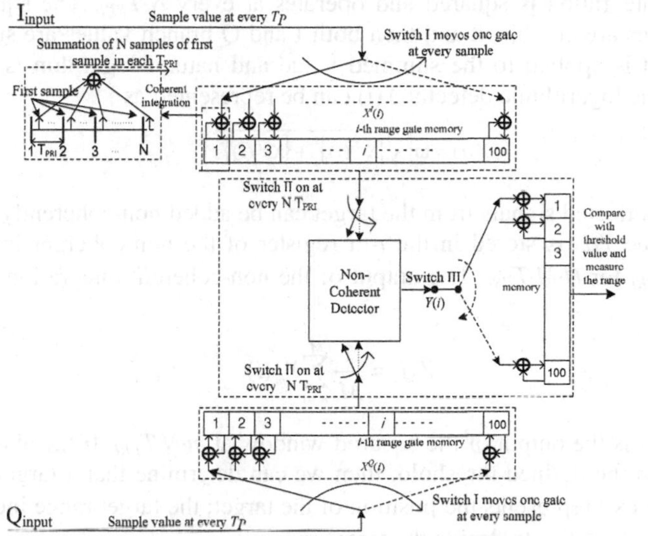

Fig. 3 Block diagram of receiver with detector

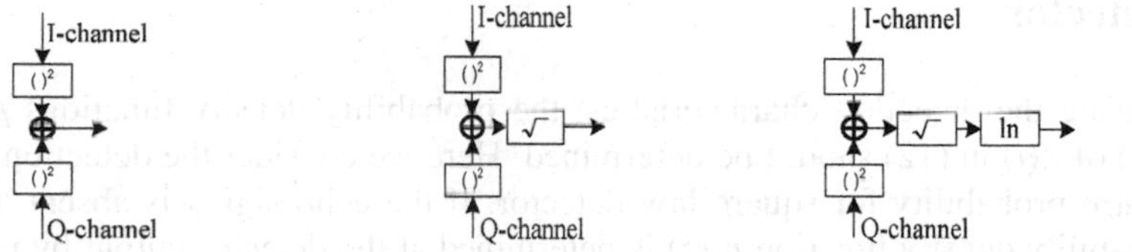

(a)Square Law (b) Linear (c) Logarithmic Detector

Fig. 4 Non-coherent Detector

The sample value received from the coherent integration is squared and operates at every $N{\cdot}T_{PRI}$. The squared range gate samples are combined and then both I and Q branch values are summed as shown in fig. 4a. The output after squaring $Y(i)$ is known as square law detector can be represented as

$$Y(i) = \{X^I(i)\}^2 + \{X^Q(i)\}^2 \tag{9}$$

In the case of a linear detector as shown in fig. 4b the sample value received from the coherent integration is squared and operates at every $N{\cdot}T_{PRI}$. The squared range gate samples are combined and then both I and Q branch values are summed and square root is applied to the summed value. The output of the linear detector $Y(i)$ can be represented as

$$Y(i) = \sqrt{\{X^I(i)\}^2 + \{X^Q(i)\}^2} \tag{10}$$

In Logarithmic detector as shown in fig. 4c the sample value received from the coherent integration is squared and operates at every $N{\cdot}T_{PRI}$. The squared range gate samples are combined and then both I and Q branch values are summed and square root is applied to the summed value and natural logarithm is taken. The output of the logarithmic detector $Y(i)$ can be represented as

$$Y(i) = \ln\left(\sqrt{\{X^I(i)\}^2 + \{X^Q(i)\}^2}\right) \tag{11}$$

All of the reflected signals from the target can be added non-coherently.

The value $Y(i)$ is stored in the i-th register of the non-coherent integration at every $N{\cdot}T_{PRI}$ for $N{\cdot}M{\cdot}T_{PRI}$. The output of the non-coherent integration $Z(i)$ can be written as

$$Z(i) = \frac{1}{M}\sum_{m=1}^{M} Y_m(i) \tag{12}$$

where $Y_m(i)$ is the output of the squared window at $m{\cdot}N{\cdot}T_{PRI}$. If the above result is greater than the defined threshold, then we can determine that a target is present. And the index i represents the position of the target; the target range indicates $i \cdot 30$ cm. It takes $N{\cdot}M{\cdot}T_{PRI}$ to decide the target range

4 Detection and False Alarm Probability for Square Law Detector

To calculate the detection characteristics, the probability density functions $p_0(z)$ and $p_1(z)$ of $Z(i)$ in (12) should be determined. Here we consider the detection and false alarm probability for square law detector. If the echo signal is absent, then the probability density function $p_0(z)$ is determined at the detector output by using the following expression [6]

$$P_0(Z) = \frac{1}{\sigma^2 2^{n/2}\Gamma(n/2)} Z^{(n/2)-1} e^{-z/2\sigma^2} \tag{13}$$

where the central chi-square distribution $p_0(z)$ with n degree of freedom has zero mean and variance σ^2. And if the echo signal is present, then the probability density function $p_1(z)$ is determined as following [6]

$$P_1(Z)=\frac{1}{2\sigma^2}\left(\frac{z}{s^2}\right)^{(n-2)/4} e^{-(s^2+z)/2\sigma^2} I_{(n/2)-1}\left(\sqrt{z}\frac{s}{\sigma^2}\right) \tag{14}$$

where $I_{n/2-1}$ is the n-th order modified Bessel function of the first kind. The non-central chi-square distribution $p_1(z)$ with n degree of freedom has mean s^2 and variance σ^2.

To get the detection and false alarm probability, we must calculate the following integral; the probability of false alarm can be written as [6]

$$P_{FA}(Z)=\int_{u_{th}}^{\infty}\frac{1}{\sigma^2 2^{n/2}\Gamma(n/2)}Z^{(n/2)-1}e^{-z/2\sigma^2}dz \tag{15}$$

and the probability of detection is given as

$$P_d(Z)=\int_{u_{th}}^{\infty}\frac{1}{2\sigma^2}\left(\frac{z}{s^2}\right)^{(n-2)/4} e^{-(s^2+z)/2\sigma^2} I_{(n/2)-1}\left(\sqrt{z}\frac{s}{\sigma^2}\right)dz \tag{16}$$

where u_{th} is the threshold value. On the basis of the above mentioned formulas, the detection characteristics of the received echo signal is determined for proposed detector.

Table 1 System parameters

Parameter	**Notation**	**Value**
Pulse Repetition Interval	T_{PRI}	200ns
Pulse Width	T_P	2ns
Maximum Target Range	R_{max}	30m
Range Resolution	R	30cm
Number of coherent integration	N	10
Number of non-coherent integration	M	variable

5 Computer Simulation Results

The purpose of the simulation is to assess the performance of the non-coherent detectors. First, we compare theoretical results with computer simulation results by using the probability density function. And then the probabilities of detection and false alarm are evaluated. In the simulations, we use the percentage of total energy reflected from each flare point as 1. We simulate the probability density functions using the system parameters as given in table I. A large enough number of trials are used to obtain each point of the probability density functions. The number of trials is about 1000000 times. The signal-to-noise ratio (SNR) is defined as $\bar{E}/N_0$, where $\bar{E}$ represents the total average energy reflected from a target.

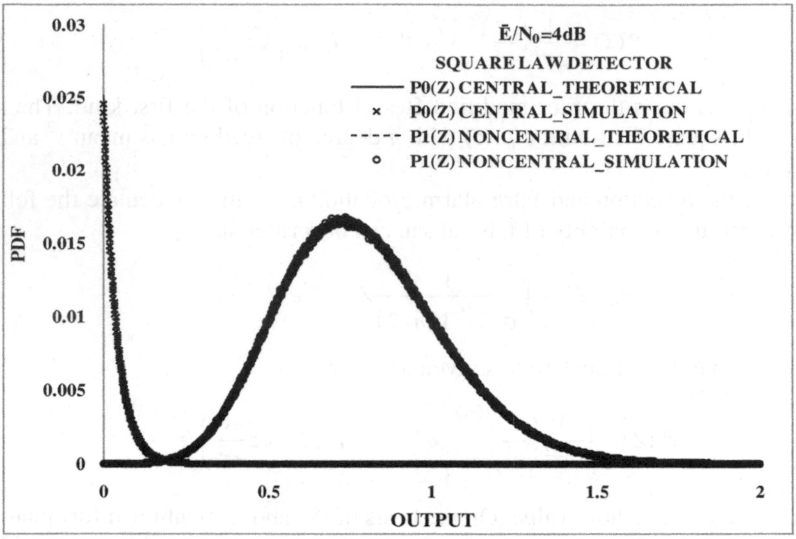

Fig. 5 The probability density function

Fig. 5 shows the result of the probability density functions (PDF) $p_0(z)$ and $p_1(z)$ $\bar{E}/N_0$=4dB for non-coherent detector. The simulation result is compared with the theoretical result. It shows that the simulation result and the theoretical result are in excellent agreement. And the PDF has 2 degrees of freedom and the average signal energy is 1. By using the probability density functions, the detection characteristics of the detector can be plotted.

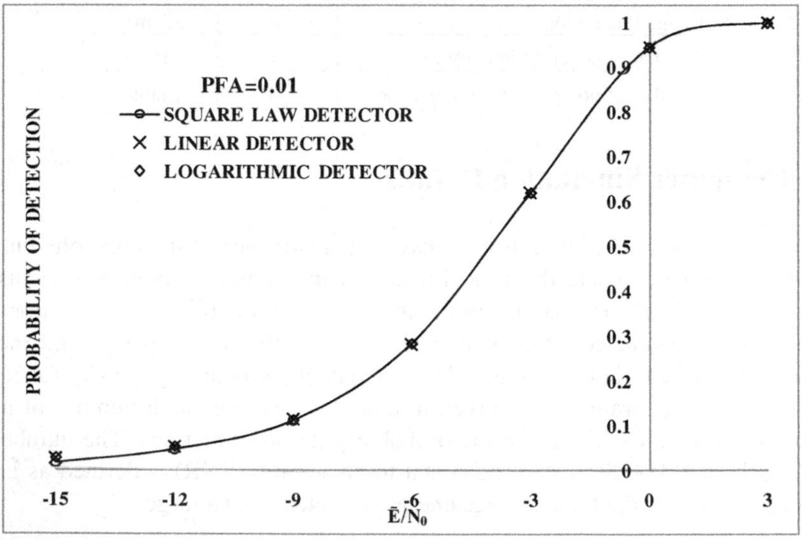

Fig. 6 The detection probability vs. $\bar{E}/N_0$ at P_{FA}=0.01

Fig. 6 shows the detection probability versus $\bar{E}/N_0$ for stationary target at P_{FA}=0.01. The non-coherent detectors have detection probability of 1 at $\bar{E}/N_0$=3dB and the performance of all the three detectors is same.

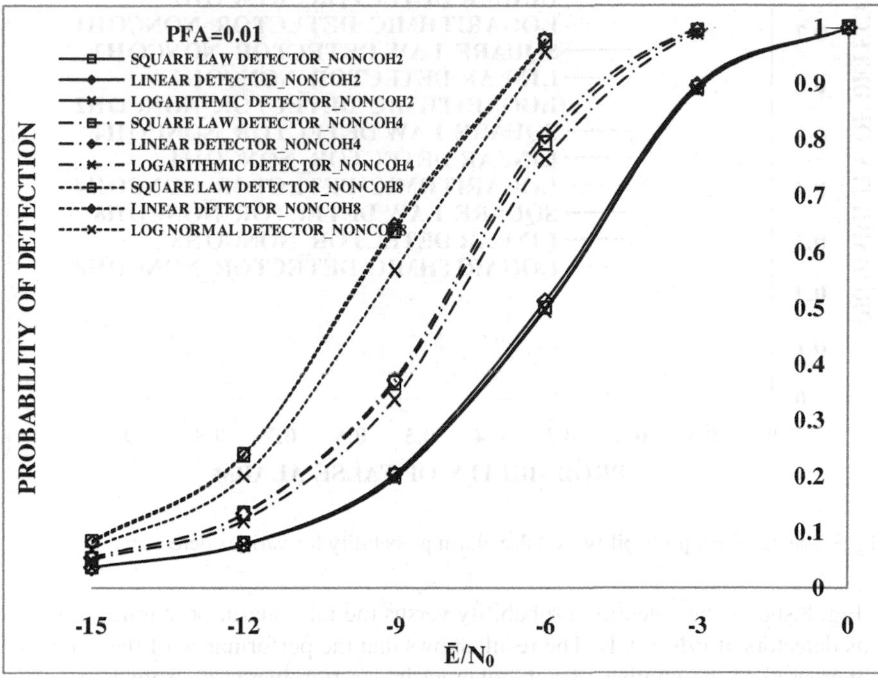

Fig. 7 The detection probability vs. $\bar{E}/N_0$ at P_{FA}=0.01

Fig. 7 shows the detection probability versus $\bar{E}/N_0$ for various stationary target at P_{FA}=0.01. The non-coherent detectors have better detection probability as the number of non-coherent integration increases. The simulation result shows that at $\bar{E}/N_0$=-6dB the probability of detection is approximately 1 for non-coherent integration number (NONCOH) of 8 for all the detectors, On the other hand the probability of detection reduces by 0.2 when the number of non-coherent integration reduces to 4.We can also predict that the performance of all the detectors is almost same when we use the same number of non-coherent integration. Therefore we can use any of the three detectors for automobile applications.

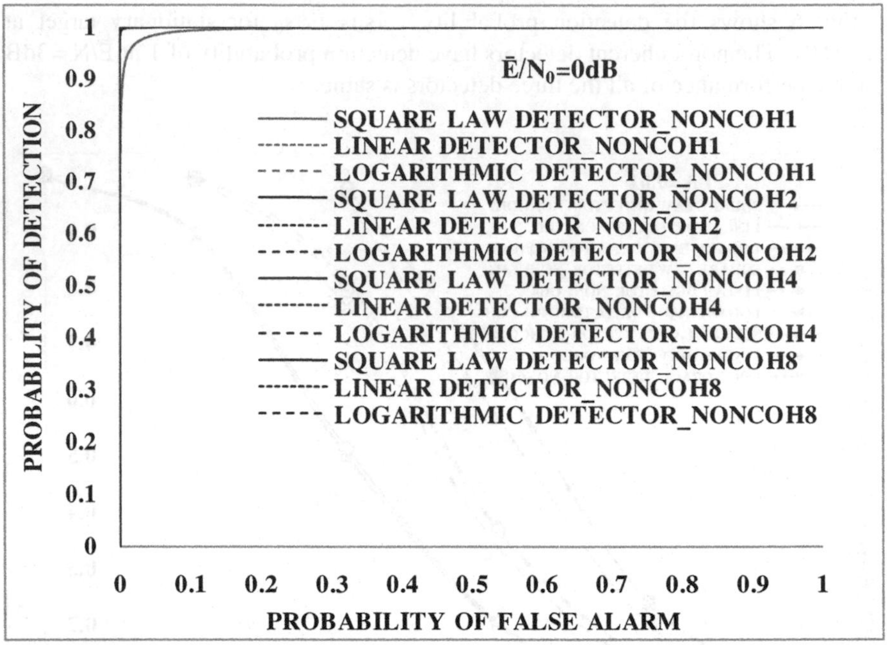

Fig. 8 The detection probability vs. false alarm probability for various detectors

Fig. 8 shows the detection probability versus the false alarm probability for various detectors at $\bar{E}/N_0$=0dB. The result shows that the performance of the detectors is increased as the number of non-coherent integration increases from 1 to 8. We can predict that all the non-coherent detectors mentioned in this paper have similar performance in a background of white Gaussian noise.

6 Conclusion

In this paper, we have analyzed the performance of non-coherent detection algorithm for Ultra Wide Band Short Range Radar (UWB-SRR) signals in automotive applications. The detection probability is found to be same for all the non-coherent detectors such as square law detector, linear detector and logarithmic detector in various SNR. Also, in order to get the detection probability to be above 0.9 for P_{FA}=0.01, $\bar{E}/N_0$ is required to be more than 0dB. Therefore it is necessary that the number of coherent integration must be increased.

Acknowledgments. This work was supported by the research grant of Jeju National University in 2007. This paper is already published in Springer, May 2011.

References

1. Strohm, K.M., et al.: Development of Future Short Range Radar Technology. In: Radar Conference (2005)
2. Taylor, J.D. (ed.): Ultra-Wideband Radars Technology: Main Features Ultra-Wideband (UWB) Radars and differences from common Narrowband Radars. CRC Press (2001)
3. Taylor, J.D. (ed.): Introduction to Ultra-Wideband (UWB) Radars systems, Boca Raton, FL (1995)
4. Hughes, P.K.: A High-Resolution Radar Detection Strategy. IEEE Transaction on Aerospace and Electronic Systems ASE-19(5), 663–667 (1983)
5. Van Der Spek, G.A.: Detection of a Distributed Target. IEEE Transaction on Aerospace and Electronic Systems ASE-7(5), 922–931 (1971)
6. Proakis, J.G.: Digital Communications. McGraw-Hill (2001)
7. Surendran, P., Ko, S.J., Kim, S.-D., Lee, J.-H.: A Novel Detection Algorithm for Ultra Wide Band Short Range Radar in Automobile Application. In: IEEE VTC 2010-Spring (May 2010)

Author Index